The Univ Equation

— a simple, new method for microcomputers

by

Noel Kantaris and Patrick f. Howden

Sigma Technical Press

ISBN 0 905104 40 4

Published by:

SIGMA TECHNICAL PRESS,
5 Alton Road,
Wilmslow,
Cheshire,
UK.

Typesetting and Production by:

Designed Publications Ltd,
8-10 Trafford Road,
Alderley Edge,
Cheshire.

Distributors:

Europe, Africa:
JOHN WILEY & SONS LIMITED,
Baffins Lane, Chichester,
West Sussex, England.

Australia, New Zealand, South-East Asia:
Jacaranda-Wiley Ltd., Jacaranda Press,
JOHN WILEY & SONS INC.,
GPO Box 859, Brisbane,
Queensland 40001, Australia.

Printed and bound in Great Britain by
J. W. Arrowsmith Ltd., Bristol

PREFACE

Books on equation solving and matrix inversion are often truly vast, containing a morass of laborious specialized techniques for solving (more often only partly solving) some of the multitude of different equation types. Consequently, the benefits to be derived from inverting equations are often lost from sight.

The new, non-recursive algorithm presented in this book will solve equations encountered in any field of endeavour, either of the simple F(x)=0 type, or simultaneous or complex types. The equations can be linear, polynomial, transcendental or differential. The roots can be positive, negative, very large or small, multiple, very close together, real or complex. Singularities are also precisely determined. The algorithm can handle poorly behaved functions, possibly containing discontinuities and infinities.

The new technique, unlike the Newton-Raphson method, does not require differentials to be found. It reaches solutions, to a predefined accuracy in very few steps and has a large capture range, without the problem of accumulating rounding-off errors.

The method uses a simple algorithm which can be manipulated by anyone capable of handling a scientific calculator without the need for guessing close trial solutions. As the method is self-correcting against introduced errors, even an unskilled operator can obtain accurate results.

This book should prove invaluable to all Engineers, Technologists, Scientists, Mathematicians, Researchers, Teachers and Students or indeed anyone whose work leads to the need for solving equations of various types rapidly and with great precision.

The examples presented here were chosen from as wide an area of human interest as possible in order to illustrate the power and versatility of the new algorithm.

A small BASIC program is provided for the use of those who prefer the versatility of a computer. The program allows the user to change initial parameters and is so flexible that it can be used to solve several sets of related or independent, single or simultaneous equations, all at once. It is equipped with subroutines which allow the user to search for closely-spaced roots, in the case of a single equation, or rotate the order of entry of simultaneous equations in order to assure automatic convergence. The program can also be used to solve for the turning points of functions and surfaces. Data fitting, splines and initial-value differential equations can be dealt with easily.

Finally, an expanded version of the program can be used (with minimum formatting requirements) to solve up to four simultaneous differential equations, each one of which can be up to the fourth differential order. The method can handle implicit differential equations with equal ease.

In Chapter 1 some traditional techniques, used to solve for the roots of a single equation, are discussed in order to establish their fallibility, before delving into the general formulation of the new algorithm. Chapter 2 deals with simple single-equation examples chosen specifically in order to illustrate the 'calculator' and 'computer' techniques applicable to the new algorithm.

As the new algorithm is equally adept at solving two or more simultaneous equations, Chapter 3 deals with the procedure to be followed in order to solve simultaneous or complex equations with the 'calculator' and 'computer' methods, while in Chapter 4 we discuss in detail the solution of several problems taken from the areas of Physics, Management, Electronics and Mathematics.

Chapter 5 deals with the problem of finding turning points of functions and surfaces, while in Chapter 6, we discuss the application of the new algorithm to such problems as interpolation, data fitting and splines.

Chapter 7 deals with the formatting procedure required in order to solve either single or simultaneous initial-value differential equations. First, the Taylor's series method of solving initial-value differential equations is discussed before showing how the same program can be used to solve any number of simultaneous differential equations of any complexity and order.

In Chapter 8, we show how to further develop the computer program so as to incorporate the required number of Taylor's series terms into it, thus substantially reducing the burden of formatting differential equations. The method can solve up to four simultaneous differential equations each one of which can be up to the fourth differential order. In Chapter 9, we discuss boundary value, Eigenvalue and partial differential equations, while in Chapter 10, we introduce techniques for injecting constraints into equations.

The computer program being evolved throughout the main body of the book was written on the Apple II microcomputer. However, a shorter version of ROOTS and a simplified version of the SIM.ROOTS programs are given in Appendix B. These MICRO versions were implemented on both the Apple II and the BBC microcomputers, as well as on the Sharp 1500 hand-held computer. They are more suitable for one-line, hand-held computers or programmable calculators which might not have some of the functions used in the versions of the program given in the main body of the text, or might not have sufficient working memory.

As the program capability grew, it became increasingly difficult to avoid machine-dependent coding, so the SIM.ROOTS version of the program was implemented also on the BBC microcomputer and a listing of it is included in Appendix C.

Although we make reference in the text to several program names, (i.e. ROOTS, SIM.ROOTS and DIFF.ROOTS), they are all versions of the same program with add-on facilities which, however, demand more coding and memory. Nevertheless, it must be pointed out that the SIM.ROOTS program will also perform all the functions of the ROOTS program (apart from the continuous root-searching mode), while the DIFF.ROOTS program will also perform all the functions of the other two programs.

If you wish to purchase a disk (DOS 3.3 for the Apple or FORMAT 40 for the BBC) containing all the versions of the Roots program together with all utility programs mentioned in this book, contact Dr Noel Kantaris at the Camborne School of Mines, Pool, Redruth, Cornwall.

ACKNOWLEDGEMENTS: We would like to thank Mary Williams, Lecturer in Mathematics at Cornwall Technical College, for her help and encouragement, Antony Campbell and Jenni Spear for checking the manuscript and for the editing suggestions they made.

LIST OF CONTENTS

CHAPTER 1 ROOT SOLVING TECHNIQUES

CHAPTER 2 FINDING THE ROOTS OF AN EQUATION

CHAPTER 3 SOLVING SIMULTANEOUS EQUATIONS

CHAPTER 4 SOLVING PRACTICAL PROBLEMS

CHAPTER 5 TURNING POINTS

CHAPTER 6 DATA FITTING

CHAPTER 7 INITIAL-VALUE DIFFERENTIAL EQUATIONS

CHAPTER 8 DIFF.ROOTS PROGRAM FOR INITIAL-VALUE EQUATIONS

CHAPTER 9 BOUNDARY, EIGENVALUE AND PARTIAL DIFFERENTIAL EQUATIONS

CHAPTER 10 CONSTRAINTS IN EQUATIONS

CHAPTER 1

Root solving techniques

In this Chapter we shall first establish the fallibility of some traditional numerical techniques in solving the general equation $F(x)=0$, prior to introducing the new algorithm. There are few, if any, restrictions imposed on this function which may be either linear, polynomial of any degree, complex, poorly behaved with possible discontinuities, or transcendental. Equally, its roots or zeros may be either single, multiple, real, complex or identical. Singularities (infinities) whose precise values could be required, may also be present. If you are not interested in the way traditional methods work or in their limitations, then skip to Section 1.3. If, on the other hand, you would like to read more about these methods, you might find it useful to read the books by Gerald (1970), Noble (1970) and Cohen (1973).

1.1 Conventional methods

All the iterative methods begin with some rough approximation, x_n, to a real root, and refining it by a systematic repetitive, non-recursive technique until the value of the root is obtained to any desired accuracy. Some other techniques, such as that of straight-line fitting, are not necessarily of the formal iterative type, but are heavily dependent on the operator's judgement in providing trial solutions.

(A) FIRST-ORDER NEWTON-RAPHSON

The first-order Newton-Raphson method states that, if x_n is the guessed root to the equation given by $F(x)=0$, then a better approximation to the root, x_{n+1} can be obtained by considering the following algorithm:

$$x_{n+1} = x_n - \frac{F(x_n)}{F'(x_n)} \qquad (1.1)$$

where $F(x_n)$ is the value of the original function, while $F'(x_n)$ is the value of the differential of the original function, both with x_n substituted into the expressions.

Referring to Fig. 1.1, suppose that R is the desired root of the equation $F(x)=0$; let x_0 be the initial approximation to the root (near enough to R) that the tangent cuts the axis at x_1.

This point of intersection is then the second approximation. The process is repeated until the root is approached within a desired accuracy.

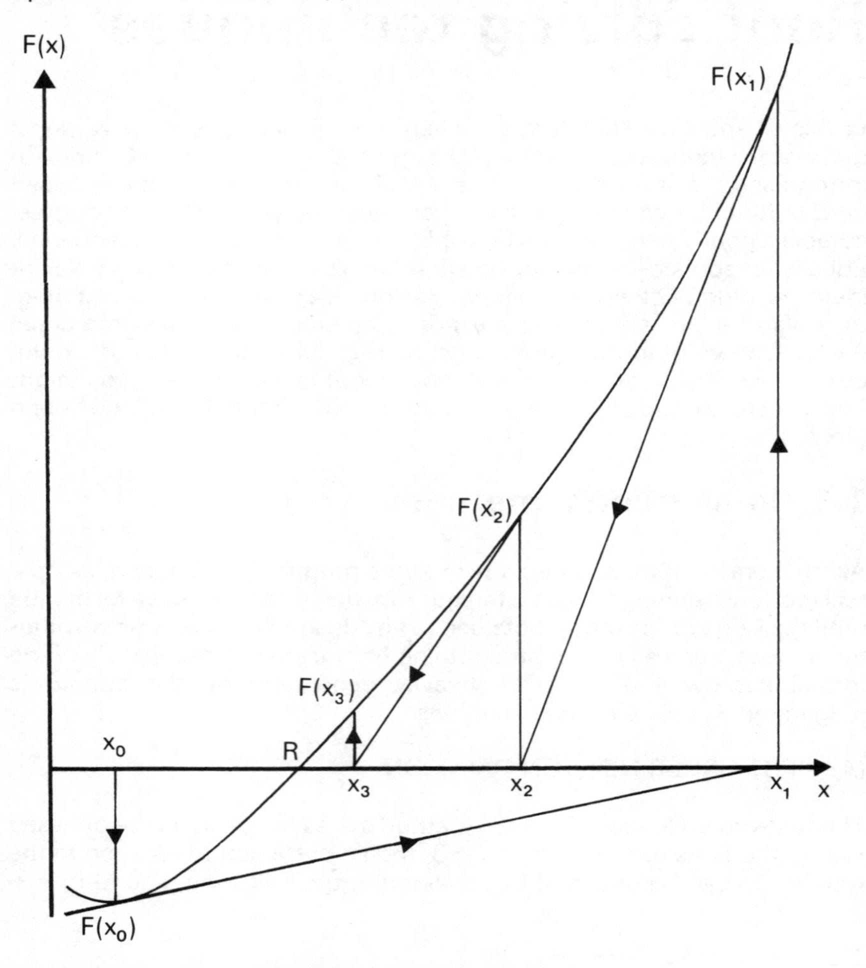

Figure 1.1 Illustration of the Newton-Raphson method.

The Newton-Raphson algorithm converges if, near the root,

$$\left| \frac{F(x)\, F''(x)}{F'(x)} \right| < 1$$

which infers that $F' \neq 0$, especially near the roots. F'' is the second differential of the function under consideration.

(B) SECOND-ORDER NEWTON-RAPHSON

A more precise scheme is the second-order Newton-Raphson method which is defined by taking another term in Taylor's expansion and is given by

$$x_{n+1} = x_n + \frac{F'}{F''} \left[-1 \pm \left\{ 1 - \frac{2\,F\,F''}{(F')^2} \right\}^{1/2} \right] \qquad (1.2)$$

Taking the square root to first order, for small values of F and considering the positive sign, Equation (1.2) reduces to Equation (1.1). Note that if F is taken to be of the form

$F = ax^2 + bx + c$

substituting in equation (1.2) yields the well known quadratic solution. Equation (1.2) gives vastly more accurate results, in fewer steps, than Equation (1.1), but more conditions are implied, such as $F'' \neq 0$ and that the function must not yield a negative square root.

(C) FUNCTIONAL ITERATION

The functional iteration technique is often called the $x=F(x)$ method, or more precisely, the $x_{n+1} = F(x_n)$, which depends on $|F'| < 1$ near a root for convergence. An arbitrary re-arrangement of the equation will sometimes enable a non-convergent configuration to converge. For example, an equation by W. B. Davies (Educ. Times, 1867) arranged as

$$x_{n+1} = \frac{480}{x_n^6} - \frac{28}{x_n^2}$$

will not converge for values of x starting at 2, while

$$x_{n+1} = \left[\frac{480}{28 + x_n^3}\right]^{1/4}$$

will converge, the two equations of course being the same.

(D) PARABOLIC INTERPOLATION

The parabolic interpolation method is usually better than the straight-line fit between two points $[x_1, F(x_1)]$, $[x_2, F(x_2)]$ chosen to lie on each side of a root. The method requires a third intermediate or mid-point at

$$(\bar{x}, \bar{F}) = \left[\frac{(x_1 + x_2)}{2}, F\left(\frac{x_1 + x_2}{2}\right)\right]$$

in order to fit a parabola through them whose root closely approximates that of the original equation. Thus,

$$x_{n+1} = \bar{x} + \frac{(x_2 - x_1)}{4[F(x_1) + F(x_2) - 2\bar{F}]}\Big\{(F(x_1) - F(x_2)) +$$

$$\left[(F(x_1) - F(x_2))^2 - 8\bar{F}(F(x_1) + F(x_2) - 2\bar{F}\right]^{1/2}\Big\} \qquad (1.3)$$

Strangely enough, greater accuracy is often achieved when the first order approximation of this formula is used, namely

$$x_{n+1} = \bar{x} - \bar{F}\,\frac{(x_2 - x_1)}{[F(x_2) - F(x_1)]} \qquad (1.4)$$

A difficulty arises when trying to decide which point to take next, given this value. One possibility is to average x_{n+1} and $\overline{x}$ in order to obtain a third point, which is used to find the new value of x_{n+1}. However, the outer two points must span the root which is a very personalized procedure involving a lot of initial guessing and rather long calculations, despite the advantage of dispensing with the need for differentials.

(E) BINARY SEARCH (SUB-DIVISION)

The binary search technique is sometimes used in stalking roots by computer. This non-differential method seeks sign changes in $F(x_n)$ as x_n is given values in progressive powers of 2.
The method is applied to a region where a root is suspected, say between 0 and 8, and then two different procedures can be followed:

Either (a) divide this region into two halves and search for a sign change in either section (0–4) or (4–8). If the sign change is in, say, the lower section, this section is halved and the procedure is repeated,

or (b) start at the lower end of the region, in this case at $x=0$, add to it a small increment, say 0.1, and test both points (0, and 0.1) for a sign change. If no sign change is found, double the increment, and so on, until a sign change occurs. Subsequent steps applied to the last such interval in which the sign changed, are identical to procedure (a).

Obviously, both procedures, although fairly easy to program, can entirely miss two closely-spaced roots, especially on the upper ranges of the function.

(F) SUM OF SERIES

The accuracy of the sum of series method is dependent on how many series terms are employed, the closeness of the guessed value to the root x, and whether, for any given equation, one runs out of differentials in the expression

$$x = x_0 - [a_1 + \tfrac{1}{2} a_2 a_1^2 + \tfrac{1}{6}(3a_2^2 - a_3)a_1^3 + \tfrac{1}{24}(15a_2^2 - 10a_2 a_3 + a_4)a_1^4 + \dots] \quad (1.5)$$

where

$$a_1 = \frac{F(x_0)}{F'(x_0)}, \qquad a_2 = \frac{F''(x_0)}{F'(x_0)}, \qquad a_3 = \frac{F'''(x_0)}{F'(x_0)}, \qquad a_4 = \frac{F^{IV}(x_0)}{F'(x\ )}$$

with the implied restriction that $F'(x_0) \neq 0$.

Obviously, the method requires a lot of arithmetic for a rather poor capture range.

There are several other, less well-known iterative schemes which we shall skip over as neither their manipulation, their ability to handle non-linear equations, nor the results obtained by their use are any better than those obtained by the use of the above-mentioned techniques. Instead, we shall solve two examples using the best of the traditional methods before pointing to possible areas of algorithm improvement.

1.2 Examples to illustrate conventional methods

As a first example, consider the equation

$$F = x^2 - 99x - 100 = 0 \tag{1.6}$$

with roots at−1, and +100.

Now $F' = 2x - 99$, and by the Newton-Raphson scheme it is apparent that the upper root will only be captured for a trial root set above 49.5, with the lower root only being captured for a trial root set below 49.5. Fig. 1.2 can be used to illustrate that the Newton-Raphson formula breaks down within the region FH because of the presence of a near-zero slope and calculated values simply oscillate or overshoot near such regions as HJ, so that fairly great precision is often needed in selecting the first (x_0) trial value for x_n.

Now consider a reasonable equation for comparison purposes, such as

$$F = x^4 - 256 = 0 \tag{1.7}$$

Starting from $x_0 = 2$, and applying the first-order Newton-Raphson method, we obtain the following iteration steps for x_{n+1}:

2, 9.5, 7.2, 5.57, 4.5477, 4.0911, 4.0028, 3.9998 and 4.0

in eight steps. Observe the rather large initial jump which results in an excessive number of iterations. On the other hand, starting from a trial $x_0 = 1$, the same solution (x=4) is reached in fourteen steps.

We can dampen the occurrence of very large iterative steps by slightly changing the Newton-Raphson algorithm, incorporating an attenuating factor of the form $\sinh^{-1}$ (see Fig. A.1 of Appendix A).

$$x_{n+1} = x_n - \sinh^{-1}\left\{\frac{F(x_n)}{F'(x_n)}\right\}$$

The solution is then reached in a reasonably short number of steps (5 and 4 steps for x values of 1 and 2, respectively), although the capture range (ability to reach a root from a distant initial approximation) for the roots could most likely be decreased. In Section 1.3 we discuss more fully how the $\sinh^{-1}$ function can be calculated.

Applying the normal, unattenuated, second-order Newton-Raphson method of Equation (1.2) to the same problem gives the following results for x_{n+1}:

2, 4.5653, 3.98664, 4.0

in only three steps, albeit at considerable complexity. Starting with $x_0 = 1$, this scheme results in a complex value for x_{n+1} which is not a root.

Applying the same attenuation function ($\sinh^{-1}$) to the second-order Newton-Raphson algorithm does not improve the number of iteration steps, despite the fact that each step results in a value of x_{n+1} which is closer to the solution than before, yet the capture range from x_0 is improved because the complex-value square root is avoided.

Putting $x_1 = 2$ and $x_2 = 5$ in Equation (1.7) and applying the reduced parabolic interpolation technique of Equation (1.4), then

$\overline{x} = 3.5$ and $\overline{F} = -105.938$,

giving the root at $x_1 = 4.02186$, which is quite an effective jump towards the correct value of 4.0.

In conclusion, the disadvantages in using traditional equation-solving techniques can be summarised as follows:

(a) Frequently a small capture range,
(b) The need to take differentials or having to deal with great arithmetic complexity,
(c) Instability arising from $F' = 0$,
(d) Non-convergence and oscillation,
(e) The inability to find close or coincident roots,
(f) Difficulty or impossibility in solving simultaneous or complex equations,
(g) Having to guess initial conditions and other high operator skills, and
(h) Inability to handle poorly behaved functions, discontinuities and infinities.

1.3 General formulation of new algorithm

The equation-solving method presented here was originally engineered on empirical lines. We postulate that an equation which can be represented as a function of x and can be written as $F(x)=0$, will have a root approximated by trial roots given by

$$x_1, x_2, \dots x_n, x_{n+1}, \dots$$

As with the functional iteration method, if

$0 = \pm F(x)$, then

$x = x \pm F(x) = f(x)$

as was discussed in Section 1.1(C). Thus, near a root, x_{n+1} could be approximated by a stability operator Q expressed as

$$x_{n+1} = x_n \pm Q\,[F(x_n)]$$

where Q approaches zero as $F(x_n)$ approaches zero (that is as x_n approaches the root). In particular, a suitable Q which does approach zero can be in the form given below:

$$x_{n+1} = x_n \pm 2^q F(x_n)$$

where q is an appropriate variable number to be derived later.

The variable q is adjusted at each iteration in order to adapt to the result of each computation - a sort of feedforward adaptive gain control (as in servo control theory). Thus, a potentially diverging or conversely go-slow tendency is counteracted by reducing or increasing q respectively by 1 before certain iterations.

If any x_n is substituted into F(x), e.g. $x_n = 0$, then $F(x_n)$ could be too large to iterate successfully. Therefore, Q must act as a well behaved attenuating function - the larger $F(x_n)$, the heavier the required attenuation. Further, the choice of operator Q must satisfy two more requirements. These are that

$Q[F(x_n)] = F(x_n)$ for small $F(x_n)$ and

$Q[-F(x_n)] = -Q[F(x_n)]$.

Thus, Q must be monotonic (continuously increasing or decreasing along an axis, but without any bumps), more or less symmetrical, and must not saturate (such as being asymptotic) or lockup (such as having discontinuities).

Two such operators were considered. The first,

$Q[F] = \tan^{-1}(F)$

appeared to be a good choice until an attempt was made to evaluate equations with large answers (x = 100, say). Although the answer was reached eventually, it was reached rather too slowly for comfort. A further slight disadvantage of this choice is that the $\tan^{-1}$ function has a limiting value of $\pm\pi/2$ radians. A strong advantage, however is that most simple scientific calculators support this function. This algorithm is better suited to hand calculators.

To overcome the slowness of the $\tan^{-1}$ function, the inverse hyperbolic sine ($\sinh^{-1}$) (see Fig.A.1 of Appendix A) was chosen as the Q function. The expression

$$\sinh^{-1}[F(x_n)] = \ln\left[F(x_n) + \left\{F(x_n)^2 + 1\right\}^{1/2}\right]$$

must be used with those calculators without the $\sinh^{-1}$ function and in any computer program (see Table A.2 of Appendix A).

To meet the overall criteria, the two forms of the algorithm are combined (2^q and $\sinh^{-1}$) as shown below:

$$x_{n+1} = x_n \pm 2^q \sinh^{-1} [F(x_n)] \tag{1.8}$$

where q can be a positive or negative integer, or zero, especially at the first step.

The above algorithm, unlike the traditional Newton-Raphson method, is completely stable and convergent under varied conditions. Although there may be other shaping functions apart from $\sinh^{-1}$, at best they could only save a few iteration steps in reaching the desired accuracy.

We have found that programming is greatly simplified by the adoption of the following algorithm as will be shown later:

$$x_{n+1} = x_n \pm 2^{\left(\frac{p}{3} - r - \frac{1}{3}\right)} \sinh^{-1} [F(x_n)] \tag{1.9}$$

where r is increased by one when the sign of $F(x_n)$ changes, and p is increased by one when the computation is going too slowly as described below.

1.4 Procedure for finding roots

The procedure to be described is related to electronic analog-to-digital conversion, whilst the 2^q multiplier is simply a crude averaging method having greater simplicity than say a parabolic or straight-line fit.

Thus, if $\pm \sinh^{-1} F(x_{n+1})$ is opposite in sign to that of the previous $\pm \sinh^{-1} F(x_n)$, r is incremented by one whilst maintaining p the same as before. When $\pm \sinh^{-1} F(x_n+1)$ has the same sign as $\pm \sinh^{-1} F(x_n)$, then p is incremented by one whilst maintaining r at its previous value. Such is the adaptive gain control.

The powers of 2 have been optimized in order to minimize the number of iteration steps taken to reach a preset accuracy in the root. Thus,

$$2^{\left(\frac{p}{3} - r - \frac{1}{3}\right)}$$

can be termed the *critical damping factor,* another term taken from servo control theory.

Figure 1.2 further illustrates how, without taking differentials, the alogrithm converges to the right on a root regardless of curve shape or distance of the trial solution from the actual root. In Fig. 1.2, distances AC, CE etc., have the relationship:

$AC = \sinh^{-1} [F(x)] = \sinh^{-1} [AB,]$

$CE = \sinh^{-1} [CD]$, etc.

Iteration towards x_R is usually much quicker because of the adaptive multiplier. Note the reversing sign change at N and corresponding halving of the increment in approaching x_R.

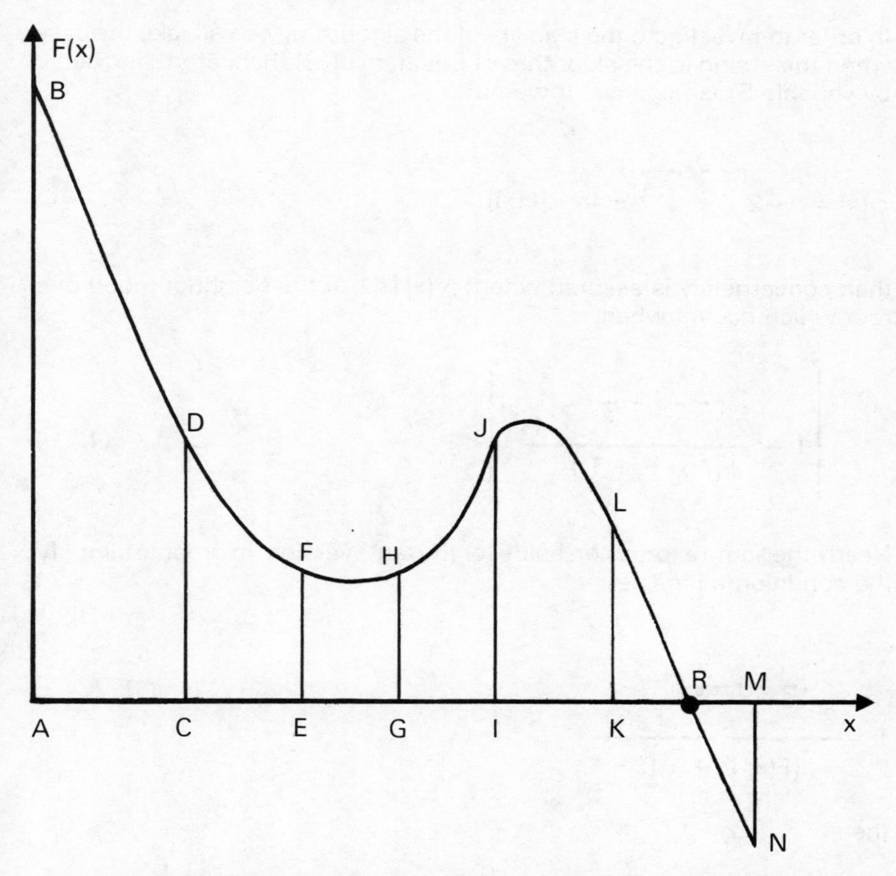

Figure 1.2 Schematic representation of the equation-solving technique.

Furthermore, reversing the sign from + to − in either Equation (1.8) or Equation (1.9) results in reversing the direction to a possible solution towards the left of the illustration. Similarly, if a new trial solution is set just to the right of the sketched x-axis crossing (point R) by a small increment, and the original + sign is also reversed, the iteration will seek further possible roots which are greater than that shown, because the function to the right of point R is negative.

Readers wishing to get straight into solving equations or whose differentiating ability is rusty, may safely skip to the next chapter.

1.5 Stability of algorithm

In order to investigate the stability of the algorithm, we will take the case when the $\pm$ sign in the algorithm of Equation (1.9) (henceforth indicated by variable S) is negative. If we put

$$\phi(x) = x - 2^{\left(\frac{p}{3} - r - \frac{1}{3}\right)} \sinh^{-1} [F(x)],$$

then convergency is assured when $|\phi'(x)| < 1$ in the neighbourhood of a root which occurs when

$$\left| 1 - \frac{2^{\left(\frac{p}{3} - r - \frac{1}{3}\right)} F'(x)}{\{ [F(x)]^2 + 1 \}^{1/2}} \right| < 1 \qquad (1.10)$$

Nearly the same expression holds for the $\tan^{-1}$ version. In order to simplify the condition a little, let

$$h = \frac{2^{\left(\frac{p}{3} - r - \frac{1}{3}\right)} F'(x)}{\{ [F(x)]^2 + 1 \}^{1/2}}$$

then

$|1 - h| < 1$ is satisfied by the condition $0 < h < 2$ and hence,

$$r - \frac{p}{3} + \frac{4}{3} > \frac{1}{\ln 2} \ln \left[\frac{F'(x)}{\{[F(x)]^2 + 1\}^{1/2}} \right] \qquad (1.11)$$

which, providing the algorithm rules given previously are obeyed, will always be true, except for one special case, namely when $F'(x)=0$, that is when coincident roots (i.e. turning points) are right on the x-axis. This one unique root type, which in real life experience will seldom, if ever, occur, will be solved if the algorithm is used to find the root of $F'(x)=0$, or by the normal algorithm method, providing there is no crossover, as will be described later.

Equation (1.11) has been checked against typical converging problems yielding reasonable consistency with a wide range of parameters. The algorithm is unconditionally stable apart from the above single case, providing also that:

$F'(x)<0$

is satisfied near a root. This is the condition on choosing a correct sign for the root in question being found.

Apart from the mathematical value of the above criteria, this is incidental and we will not use or refer to it again in the text.

To illustrate the convergency advantage of the new algorithm, relative to the independent use of p, r and $\sinh^{-1}$ operators on a straight-line function such as

$$F = m(x - 20), \qquad (1.12)$$

where m is the slope of the line, we refer the reader to Figs 1.3 and 1.4, where the equation has been solved by the functional iteration technique, as well as the $\sinh^{-1}$ algorithm.

Ignoring the ⅓ term in the exponent of the algorithms and choosing $S=-1$, the iterative algorithms required are:

$$x_{n+1} = x_n - 2^{\left(\frac{p}{3} - r\right)} m(x_n - 20) \qquad (1.13)$$

and

$$x_{n+1} = x_n - 2^{(\frac{p}{3} - r)} \sinh^{-1}[m(x_n - 20)] \tag{1.14}$$

These algorithms are compared in their ability to converge on the root $x=20$, to within an accuracy of ± 0.0001, with or without the use of p and r, at various slopes of the line. In every case, the starting condition is $x_0=0$ and the capture range is always 20.

Fig. 1.3(a) shows a plot of Equation (1.12) with the bottom half of the figure depicting the iterative steps of algorithm (1.13), with $p=0$, $r=0$ and $m=1000$. The starting and all subsequent iterations are shown by a +sign, except when $0 > x_{n+1} > 30$.In this particular case the iterations diverge rapidly causing an overflow on the fifth step.

Fig. 1.3(b) shows the iterative steps (without the plotted line) of algorithm (1.14) with the same values of p, r and m as for the previous case. It can be seen that applying the $\sinh^{-1}$ operator stops divergency, but results in a small, continuous, oscillation. In contrast to Fig. 1.3(b), Fig. 1.3(c) shows the effect of using r thereby causing a rapid, almost critically damped, convergency in 16 steps.

Fig. 1.4(a) and (b), as opposed to the previous three cases, have very shallow slopes ($m=0.1$). In both these plots, of algorithm (1.14), r is inactive, but the effect of p which speeds up convergency while the function is not changing sign, is well illustrated. Thus, we can think of the convergency in Fig 1.4(a) as being 'underdamped', while that of Fig. 1.4(b) as being 'critically damped'. Fig. 1.4(c) shows the plot of the new algorithm, as given by Equation 1.9.

Finally, Table 1.1 summarizes the effect of various options of p and r operators as applied to the algorithms (1.13) and (1.14), while the last column shows the stability and rapid convergency of the new algorithm for all slopes. For this precise reason, the new algorithm is capable of evaluating singularities.

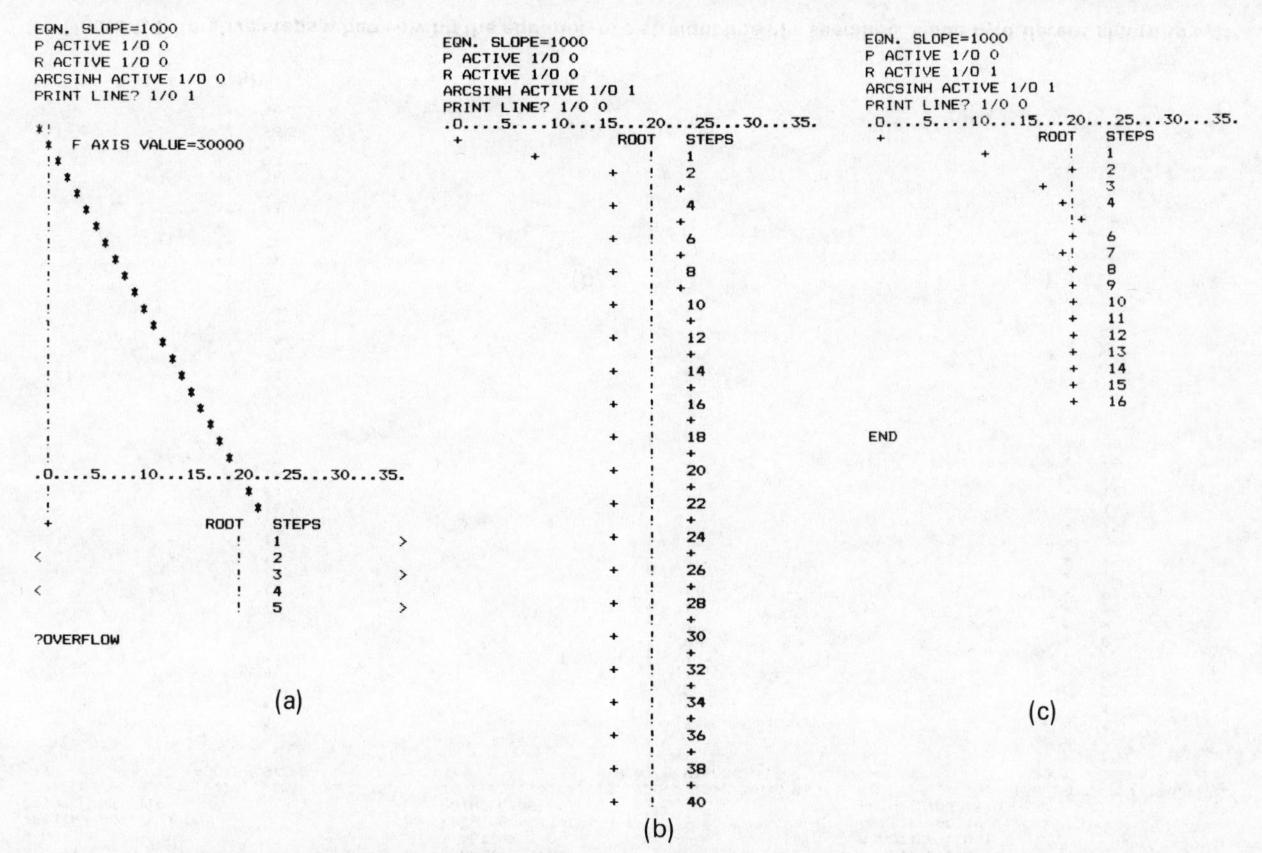

Figure 1.3 Iterative steps when solving the equation of a straight line of a specified slope by different alogorithms.

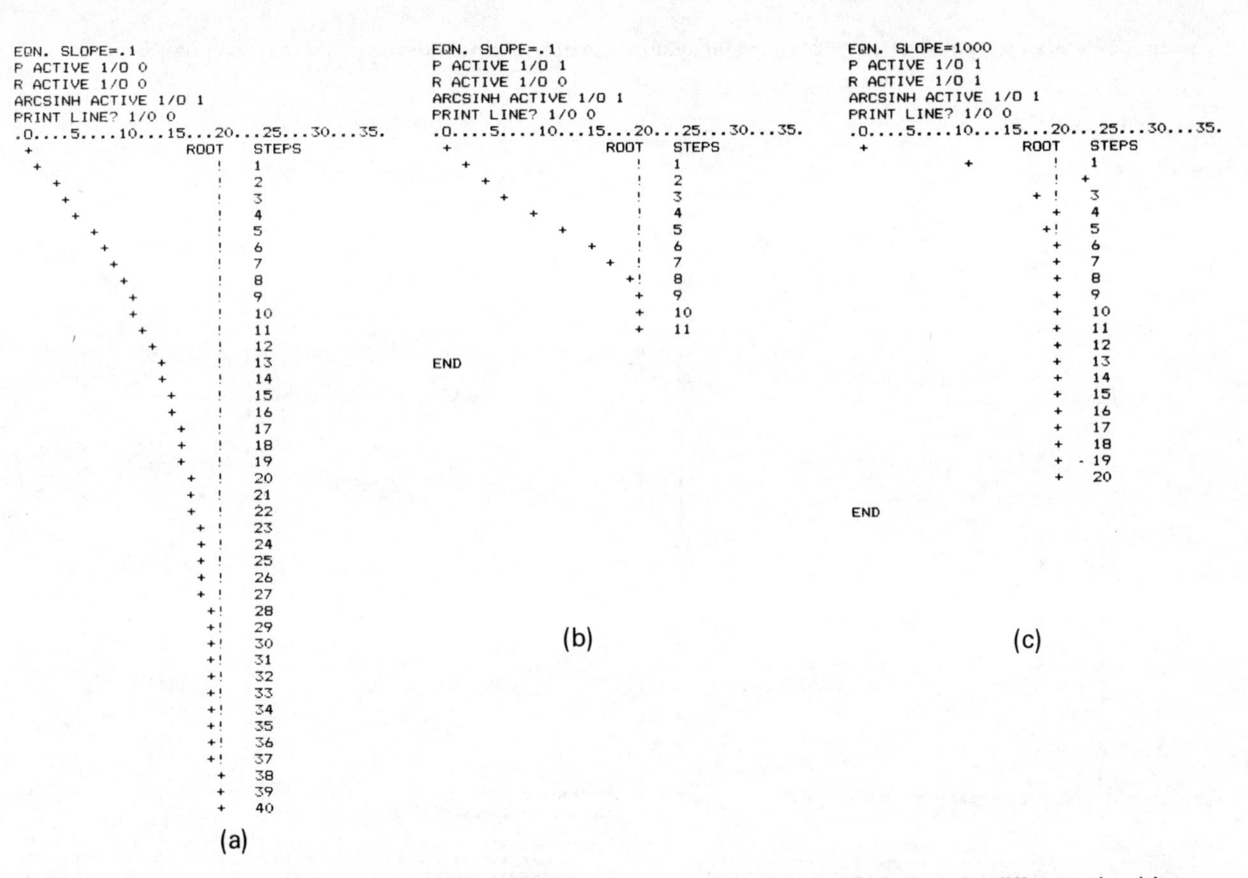

Figure 1.4 Iterative steps when solving the equation of a straight line of a specified slope by different algorithms.

TABLE 1.1 Table of Options Using the Functional Iteration or $\sinh^{-1}$ attenuator Method in Order to Converge on the Zero of a Straight Line Function, Versus Equation Slope

SLOPE OF LINE EQUAT.	NOT USING P				USING P			
	NOT USING R		USING R		NOT USING R		USING R	
	LINEAR FUNCTIONAL	$SINH^{-1}$ ATTENUATOR	LINEAR FUNCTIONAL	$SINH^{-1}$ ATTENUATOR	LINEAR FUNCTIONAL	$SINH^{-1}$ ATTENUATOR	LINEAR FUNCTIONAL	$SINH^{-1}$ * ATTENUATOR
0.01	>200	>200	>200	>200	21 No overshit	21	23	24
0.1	94	97	94	97	11	11	13	14
1	1 ∵ of Symmetry	9 No overshit	1 ∵ of Symmetry	9 No overshit	10	Small Oscillations	4 Sm overshit	9
2	Oscillates $\because \lvert F' \rvert \geqslant 1$	Small Oscillations	2 ∵ of Symmetry	8 No Overshit	Diverges	↓	5	10
10	Oscillates & Diverges	↓	19 Lg overshit	12 No overshit	↓	↓	11 Overshoot	11 No overshit
100	↓	↓	26 Lg overshit	14 Sm overshit	↓	↓	21 Lg overshit	19 Sm overshit
1000	↓	↓	Diverges	16 Sm overshit	↓	↓	Diverges	20 Sm overshit
10000	Diverges	Small Oscillations	Diverges Faster	24 Sm overshit	Diverges Faster	Large Oscillations	Diverges	36 Sm overshit

* *New algorithm as used in equation-solving program.*
Numbers in Table indicate the iterations required to reach the root to within the decimal accuracy of .0001.

CHAPTER 2

FINDING THE ROOTS OF AN EQUATION

In this Chapter we shall first use the algorithm to solve for the roots of a simple equation using the calculator procedure.

Although the method is equally adept to solving for the roots of extremely complicated equations, we will solve such equations using a small computer program which will be introduced in Section 2.3. In a later chapter, we will expand both methods so that simultaneous equations can be solved. If you are only interested in the computer method, then skip to Section 2.3.

2.1 Using a calculator to find a root

We will use the algorithm of Equation (1.9) in order to home in on roots of single equations. The algorithm can be written as:

$$x_{n+1} = x_n + 2^q H,$$

where $H = S \sinh^{-1}[F(x_n)]$ and S is the $\pm$ sign inside the curly brackets of the full algorithm expression, as shown below:

$$x_{n+1} = x_n + \left\{ \pm 2^{\left(\frac{p}{3} - r - \frac{1}{3}\right)} \sinh^{-1}[F(x_n)] \right\} \qquad (2.1)$$

The function $\sinh^{-1}$ acts as a crude attenuator, while the term 2^q behaves as the critical damping factor which is adjusted according to the iteration cross-over signs. The $\tan^{-1}$ function is often easier to use on small calculators which don't have a $\sinh^{-1}$ button.

In order to illustrate the procedure to be followed, we will solve the following equation set correctly with the function entirely on the left-hand side:

$$F(x_n) = x_n \log_{10} x_n - 5 = 0.$$

We first choose the initial conditions which are fairly arbitrary except that in this case we cannot set $x_0 = 0$ because of the $\log_{10}$ term in the function.

Usually r=0 initially unless you are expecting close roots in which case you should make it equal to 5 or 10 as will be explained later. Thus, with $x_0=1$, r=0, p=0, S=−1 and assuming an initial pre-sign to H (always taken as +1), we can then form Table 2.1, as shown below, by substituting the initial root ($x_0=1$) into the equation to yield $F(x_0)$ and then evaluating the various parts of the algorithm as given by Equation (2.1). Note that had we taken S=1 we would be solving for a possible different root.

There is one rule that should always be adhered to with respect to the choice of r and p. This is:

If $H(x_{n+1})$ has the same sign as $H(x_n)$ then keep r the same as before, but increment p by 1. Otherwise, increment r by 1 (indicating a cross-over), maintaining the latest value of p.

TABLE 2.1 Tabulation of Calculator Procedure

$F(x) = x \log_{10} x - 5$

x_n	$F(x_n)$	$H(x_n)$ Pre-sign=+1	Is SIGN($H(x_n)$) = SIGN($H(x_{n-1})$)	New p	New r	x_{n+1}
1.000	-5.00	+2.312	Yes increase p	1	0	3.312
3.312	-3.277	+1.902	Yes increase p	2	0	5.709
5.709	-6.801E-1	+6.363E-1	Yes increase p	3	0	6.719
6.719	+5.594E-1	-5.337E-1	No increase r	3	1	6.296
6.296	+3.091E-2	-3.090E-2	Yes increase p	4	1	6.265
6.265	-7.173E-3	+7.173E-3	No increase r	4	2	6.268
6.268	-2.757E-3	+2.757E-3	Yes increase p	5	2	6.270
6.270	-6.180E-4	+6.180E-3	Yes increase p	6	2	6.271

The first row of Table 2.1 is obtained by substituting the initial value for x_n ($x_0=1$) into the equation in order to obtain the initial value for $F(x_n)$, which in this case is -5.00. Following this, $H=S \sinh^{-1}[F(x_n)]$ is evaluated as $(-1)(-2.312)=+2.312$. This sign of H is positive, the same as the assumed H pre-sign. Thus, increment p by 1 and then calculate the new value of x_{n+1}. The procedure is continued down the table until the last two values x_{n+1} are identical to within the required accuracy.

2.2 Simplified tabulation procedure

Before we attempt another example, it is worth pointing out that most of the columns shown in Table 2.1, are not necessary as their values can be carried in the calculator. In fact, the experienced operator need only record x_n, the sign of $H(x_n)$ and the values of p and r in order to find any particular root.

The simplified tabulation procedure will be illustrated by applying it to the financial problem given below.

Problem:

If the amount A of a principal P is placed at a rate of interest R compounded x times a year, then the total interest I built up over n years is given by

$$I = A - P = P\left\{\left(1+\frac{R}{100x}\right)^{nx} - 1\right\}$$

where n is the number of years of investment.

With $n=8$, $R=10\%$ and $P=100$, find the minimum number of times a year (x), that interest must be compounded in order that the total interest due is 120.07.

Solution:

Substituting the values into the formula for I the equation becomes

$$F(x) = 100\left(1 + \frac{0.1}{x}\right)^{8x} - 220.07 = 0.$$

Using the initial conditions $x_0=1$, $p=0$, $r=0$ and $S=-1$ in Equation (2.1) we obtain the table shown below. Again, note that the initial presign of H is assumed positive.

x_n	Sign $H(x_n)$	p	r
1	+ve	1	0
3.443	+ve	2	0
3.491	+ve	3	0
3.498	−ve	3	1
3.497			

Recognising that x should be an integer number, we can stop the iteration at that point and take the answer to the problem as being x=4.

The problems given so far were intentionally made simple merely to illustrate the procedure, and in no way reflect the degree of complexity that can be handled with a calculator.

2.3 Computer method for finding roots

The BASIC program given in Listing 2.1 can be used to find the roots of a single equation which must be typed into the computer as a BASIC expression in line 10. The $\sinh^{-1}$ part of the new algorithm, as given by Equation 2.1, appears in line 35, while the rest appears in the second statement of line 50.

The program was implemented on the Apple II microcomputer. It has also been implemented on the BBC microcomputer by replacing all semicolons in the INPUT statements with commas and the function LOG() by LN(), and on the Sharp 1500 hand-held computer by changing the function LOG() to LN(), inserting the word LET before every assignment statement occurring within an IF statement, changing the first PRINT statement in line

50 to PAUSE and changing all TABs to CURSORs.

In Appendix B, we include a simpler version of this program which avoids the use of subroutines and is well suited to hand-held computers or programmable calculators.

```
 1  PRINT:PRINT "ENTER YOUR EQUATION ON LINE 10 AS:-":
    PRINT "10 F=X↑2+3*X-10:RETURN"
 5  PRINT: INPUT "ACCUR=";D:INPUT "NR OF ITER=";N:INPUT
    "X=";X: INPUT "R=";R1:INPUT "SEARCH INC=";L:INPUT
    "SIGN=";S:GOTO 20
20  PRINT:P=0:R=R1:S1=1:D1=0:K1=0:K2=0:IF L<>0 THEN K1=1:
    K2=1:PRINT "F(X)";TAB(20);"X":PRINT:GOTO 30
25  PRINT "ITER";TAB(8);"P";TAB(14);"R";TAB(20);"X":PRINT
30  FOR I=1 TO N
35  GOSUB 10:H=S * LOG(ABS(F)+SQR(F*F+1)) * SGN(F):S2=SGN(H)
40  IF K1=1 THEN GOSUB 65:GOTO 35
45  IF S2*S1>0 THEN P=P+1:R=R-1
50  R=R+1:X1=X+H*2↑(P/3-R-1/3):PRINT I;TAB(8);P;TAB(14);R;
    TAB(20);X1:IF ABS(X1-X)<D THEN D1=1:PRINT:PRINT "F=";:
    GOSUB 10:PRINT F;TAB(20);"X=";X1:IF L=0 THEN GOTO 5
55  IF L<>0 AND D1=1 THEN S=-1*S:X=X1+100*D*SGN(L):GOTO
    20
60  X=X1:S1=S2:NEXT I:PRINT "NOT CONVERGING IN ";N;" ITERS":
    GOTO 5
65  IF K2=1 THEN K2=0:S1=S2:GOTO 75
70  PRINT F;TAB(20);X:IF S1*S2<=0 THEN K1=0:PRINT:
    PRINT "ITER";TAB(8);"P";TAB(14);"R";TAB(20);"X":PRINT:RETURN
75  X=X+L:RETURN
```

Listing 2.1 Roots program for solving single equations

On typing 'RUN' the following message will appear on the screen:

ENTER YOUR EQUATION ON LINE 10 AS:-

10 F = X↑2+3*X-10:RETURN

ACCUR=

At this stage, press the RESET key (ESCAPE key on the BBC) and enter your equation. Note that line 10 is a subroutine and as such it must be terminated with the RETURN statement. For more complicated equations you will need to refer to the tables of Appendix A for standard BASIC and derived mathematical functions.

Having entered your equation, type 'RUN' again and provide initial values for ACCURacy, NR (number) OF ITERations, X, R, SEARCH INCrement and SIGN.

In the first instance, we ask you to answer the question 'SEARCH INC' with 0 (zero). The mode of the program obtained if you provide a non-zero answer, will be explained later.

Problem:

As a demonstration we will solve for the real roots of the following polynomial:

$F(x) = x^7 + 28\ x^4 - 480 = 0.$

Solution:

First, LOAD the program into the computer and type

```
10 F = X↑7 + 28*X↑4 - 480:RETURN
```

then 'RUN' the program and provide typical initial parameters, as shown below, to obtain:

```
ACCUR = 0.000001
NR OF ITER = 50
X = 0
R = 0
SEARCH INC = 0
SIGN = -1
```

ITER	P	R	X
1	1	0	6.86693437
2	1	1	-·264311843
3	1	2	1.45235061
4	2	2	3.50833675
5	2	3	1.94392033
6	3	3	1.16982692
7	3	4	1.83893744
8	4	4	2.48621473

9	4	5	2.00097258
10	5	5	1.58590268
11	5	6	1.83471791
12	6	6	2.0938123
13	6	7	1.94124906
14	7	7	1.82366215
15	7	8	1.90694411
16	8	8	1.97754697
17	8	9	1.92941126
18	9	9	1.89567883
19	9	10	1.92115805
20	10	10	1.93237257
21	10	11	1.92028331
22	10	12	1.92382471
23	10	13	1.92290797
24	11	13	1.92287427
25	11	14	1.92288126
26	12	14	1.92288384
27	13	14	1.92288419

F=-3.60369683E-04 X=1.92288419

indicating a root given by X. F is the value of the function when X is substituted into it. After obtaining the above result, program execution is sent to that part of the program which displays the question

ACCUR=

in order to allow a re-start.

If we change the initial conditions for R and SIGN by making them 10 and +1 respectively, while ensuring that all other input variables remain unchanged, we obtain another root

F=1.23620033E-04 X= -2.4580896

Figure 2.1 is a sketch of the function $x^7 + 28x^4 - 480$, showing the roots which can be found by appropriate choice of initial conditions.

In order to solve for the complex roots, see Section 3.4.

With SIGN=-1 and x located anywhere from a point slightly greater than the -2.45 root to a large positive value, root 1.92 will be captured.

Changing the SIGN to +1 and starting from the origin, say, results (in this particular case) in searching to the left of the origin. However, unless R is increased to 10, say, closely-spaced roots can be easily missed. Having thus found the 2.45 root, we could now increase X by a small amount (negatively) and reverse the SIGN again in order to locate the second root of the pair.

A mechanism for searching for closely-spaced roots is provided in the program. The facility, which we shall refer to as the 'Fine Search' mode, can be activated by answering the question 'SEARCH INC=' by a non-zero value. A flowchart of the program, including the 'Fine Search' subroutine, is shown in Fig. 2.2.

The fine search mode option hunts for a sign change in the value of F(x) (printed under the first column), starting from x and increasing or decreasing by a small increment L, which is an input parameter to the program. Having found a sign change, the normal ROOTS algorithm is

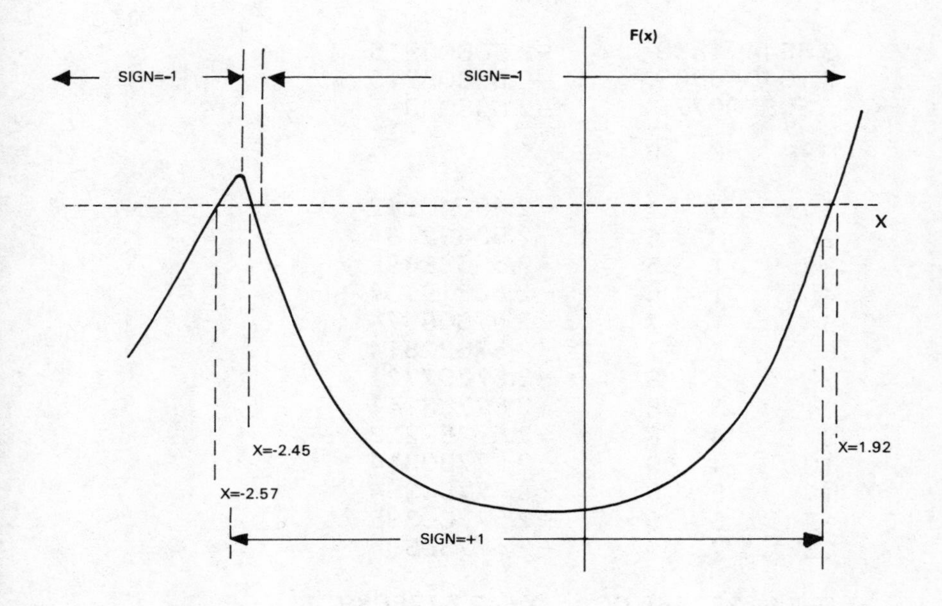

Figure 2.1 A sketch of the function $x^7 + 28x^4 - 480$.

automatically entered. Furthermore the 'Fine Search' mode will automatically search for the next root by carrying out the manual operations suggested above.

To see the 'Fine Search' mode under operation, RUN the program and provide the following initial parameters:

ACCUR = .0000001
NR OR ITER = 50
X = 3
R = 3
SEARCH INC = -.05
SIGN = -1

All three roots are found one after the other. We only show below what you will see on the screen after the second root, at X=-2.4580896, is revealed. The computer can be temporarily halted in its program execution by pressing 'CRTL'S (and re-starting it with another 'CTRL'S). For the BBC use 'CTRL'SHIFT.

F(X)	X
3.664401899	-2.50809915
2.184048889	-2.55809915
-5.30850077	-2.60809915

ITER	P	R	X
1	0	4	-2.49047212
2	0	5	-2.53492194
3	1	5	-2.59688886
4	1	6	-2.56819107
5	1	7	-2.57605574
6	2	7	-2.57832814
7	2	8	-2.57797721
8	3	8	-2.57783141
9	4	8	-2.57780228
10	4	9	-2.57780314
11	5	9	-2.57780364
12	6	9	-2.57780385
13	7	9	-2.5778039

F=5.6028366E-06 X=-2.5778039

At this point press the RESET key (ESCAPE for the BBC) as there are no further real roots to be found.

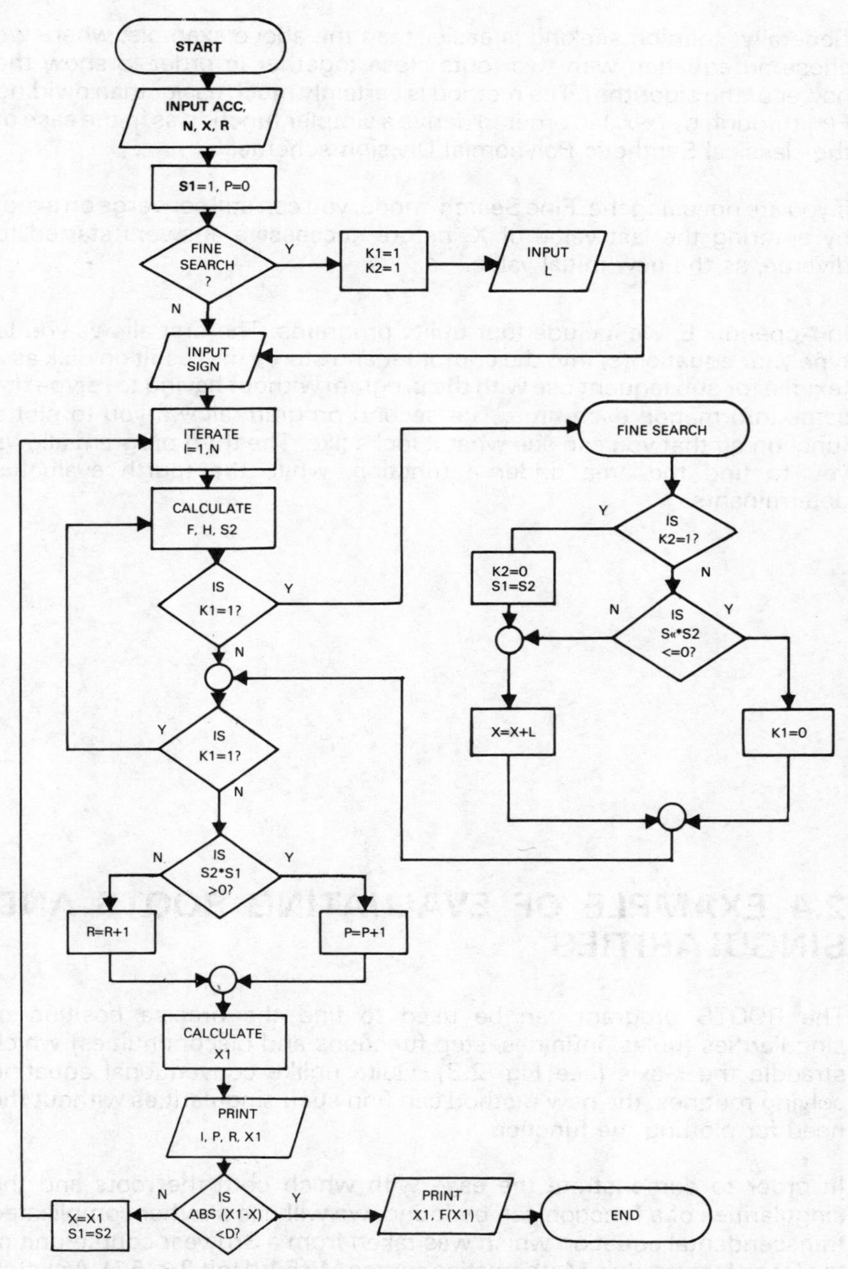

Figure 2.2 Flowchart of ROOTS program with Fine Search mode.

Generally, solution seeking is easier than the above example, where we chose an equation with two roots close together in order to show the power of the algorithm. The method is certainly much easier than dividing F(x) through by $(x-x_0)$ in order to derive a simpler function as in the case of the classical Synthetic Polynomial Division scheme.

If you are not using the 'Fine Search' mode, you can still converge on a root by entering the last value of X, before successive answers started to diverge, as the new initial value.

In Appendix E, we include four utility programs. The first allows you to type your equation(s) into the comptuer and stores the result on disk as a text file for subsequent use with the program without having to retype the same information each time. The second program allows you to plot a function so that you can see what it looks like. The third program allows you to find the area under a function, while the fourth evaluates determinants.

2.4 EXAMPLE OF EVALUATING ROOTS AND SINGULARITIES

The ROOTS program can be used to find the precise position of singularities (poles, infinities, step functions and discontinuities) which straddle the x-axis (see Fig. 2.3). Quite unlike conventional equation solving routines, the new method can find such singularities without the need for plotting the function.

In order to demonstrate the ease with which both the roots and the singularities of a function can be found, we will use a rather complicated transcendental equation which was taken from a 3rd year course-unit of the Open University's Mathematics course M351 (Unit 2, p.51). A sketch of the equation is shown in Fig. 2.3.

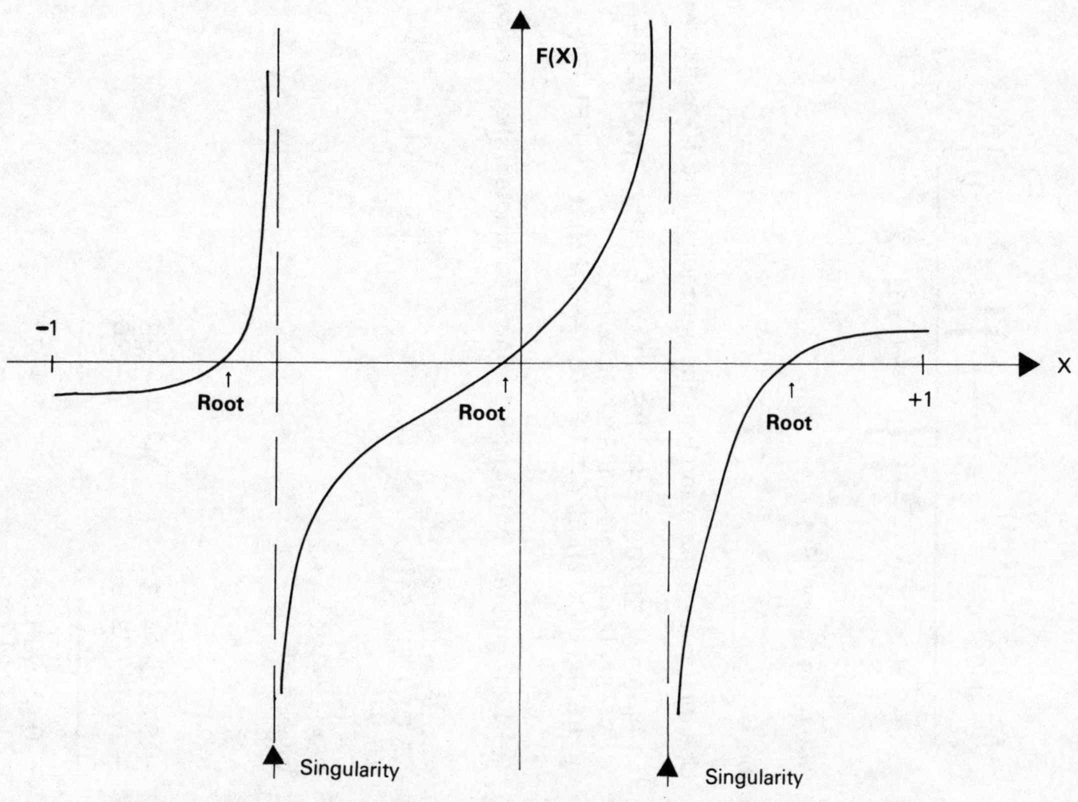

Figure 2.3 Sketch of the complicated transcendental equation clearly showing the three roots and two singularites that straddle the x-axis.

Problem:

Find all roots and singularities in

$$F(x)=\frac{\left\{\frac{\pi}{8}\sin 5x-e^{\frac{\pi}{8}(x+1)^{1/2}}+\left(x+\frac{5\pi}{8}\right)^{1/2}\right\}\left\{1+\frac{\pi}{8}(x-1)^{2}\right\}^{1/2}}{\left\{e^{-x^{2}}-\frac{\pi}{4}\right\}\left\{1+\frac{\pi}{8}(x+1)^{2}\right\}}+\frac{\pi}{8}=0$$

within the region $-1 < x < +1$.

Solution:

Use the roots program and type the equation in line 10 as follows:

```
10 PI=3.14159265:F=(PI/8*SIN(5*X)-EXP(PI/8*(X+1)↑.5)+
   (X+5*PI/8)↑.5)*(1+PI/8*(X-1)↑2)↑.5/
   ((EXP(-1*(X↑2))-PI/4)*(1+PI/8*(X+1)↑2))+PI/8:RETURN
```

'RUN' the program and provide the initial parameters shown below:

```
ACCUR = .0000001
NR OF ITER = 100
X = -1
R = 5
SEARCH INC = .02
SIGN = -1
```

F(X)	X
-.875199423	-.98
-.878570279	-.96
————	——
-.042408329	-.72
.214371175	-.7

ITER	P	R	X
1	0	6	-.702638587

2	1	6	-.705395298
—	—	—	————————
14	12	7	-.7163967

F=-1.27824E-07 X=-.7163967 ← Root

F(X)	X
.266774115	-.6963867
.596641465	-.6763867
————————	————————
115.837589	-.4963867
-39.9720978	-.4763867

ITER	P	R	X
1	0	6	-.530723999
2	0	7	-.510699356
—	—	—	————————
37	11	31	-.491492038

F=20117396.3 X=-.491492038 ←Singularity because of large F

F(X)	X
-30.5936727	-.471482038
-16.0863635	-.451482038
————————	————————
-.018283292	-.011482038
.147829854	8.5179618E-03

ITER	P	R	X
1	0	6	6.69125205E-03
2	1	6	4.6202306E-03
—	—	—	————————
16	13	8	-9.319975E-03

F=4.65661287E-08 X=-9.319975E-03 ← Root

F(X)	X
.165471527	.010690025
.324396142	.030690025
———	———
182.227463	.490690025
-5.8890171	.510690025

ITER	P	R	X
1	0	6	.480016374
2	0	7	.500800753
—	—	—	———
49	20	34	.491492205

F=-804459.456 X=.491492205 ← Singularity

F(X)	X
– 5.58250938	.511502205
– 1.95921306	.531502205
–.751467953	.551502205
–.150539593	.571502205
.205597648	.591502205

ITER	P	R	X
1	0	6	.588970101
2	1	6	.586343286
—	—	—	———
12	10	7	.578852094

F=–6.27711415E-07 X=.578852094 ← Root.

After $x > 1$, stop the computer by pressing the RESET key, as all the roots and singularities within the region $-1 < x < +1$ have been found. In order to solve for the complex roots, see Section 3.4.

If your computer does not accept the number of characters per line required to enter the complete equation in one line, then split the equation in appropriate places and enter them on different lines. However, you must observe the convention that the first line of the group must be line 10, while the last statement of the last line must be the RETURN. For example, we could enter the function as follows:

```
10 PI=3.14159265
11 A = numerator of expression
12 B = denominator of expression
13 F=A/B+PI/8:RETURN
```

Care must be taken in bracketing terms correctly. For example

$$e^{-x^2}$$

should be entered as EXP(-1*(X↑2)) and not as EXP(-X↑2) because the latter expression will always result in a positive answer and is the inverse of the desired function.

For greater stability, where an equation has close poles and roots, use $\tan^{-1}F$ which has a limiting value of $\pm \pi/2$. This will reduce the value of the function near a pole so as not to diverge from the root. To implement this, change line 13 of the above example to

```
13 F=ATN(A/B+PI/8):RETURN
```

It must be pointed out that unlike other classical methods, the new algorithm can accept poorly behaved equations containing discontinuities (steps and gaps), infinities, or such non-differentiable functions as ABS(), INT(), SGN(), etc.

2.5 SPECIAL CASES?

There are two cases that may be considered as special and merit closer look. These are:

(a) when a turning point $dF(x)/dx = 0$ sits exactly on the x-axis,
(b) when the inverse turning point $dx/dF(x) = 0$ crosses the x-axis at $F(x)=0$.

The above cases can be investigated more closely by considering a circle of radius 2 cm which

(a) touches the x-axis at x=5 cm and is described by the function

$$F(x) = 2 - \left\{ 4 - (x-5)^2 \right\}^{1/2} \quad \text{and}$$

(b) has its diameter coincident with the x-axis with the centre of the circle at x=5 cm and is described by the function

$$F(x) = \left\{ 4 - (x-5)^2 \right\}^{1/2}$$

Were we to try to solve for x for either of the above functions, with a starting value for x outside the function itself, we would encounter an immediate problem with 'Illegal Quantity Error' (in this case a negative square root) in line 10. However, the problem disappears if we were to start with an x value within the bounds of the function (in this case 3 to 7), and a sufficiently large value of R. It is worth pointing out that as the turning point is only approached from one side, the value of R never changes.

The above problem with 'Illegal Quantity Error' can be used, in fact, to locate the approximate bounds of a complicated function, or in the case of (b), above, the roots of that function. We start with a value of x which lies within the bounds of the function and work outwards, with sufficiently large value of R, say 10. Once the latest iterated value of x causes an 'Illegal Quantity Error' to occur, re-run the program with a starting value of x equal to the second last value of x from the previous run, using the same value of R, and so on, in order to reach greater precision.

Finally, roots of implicit functions, represented by

$$Z = F(x,y) = 0,$$

e.g. $F = x^5 + 2xy + y^5 - 3 = 0$ in which the function y cannot be expressed explicitly as $y = f(x)$, can be extracted by first setting y=0 and solving for

$$F = F(x,0) = 0.$$

Thus, for the above example we would solve $F = x^5 - 3 = 0$ in the usual way.

CHAPTER 3

SOLVING SIMULTANEOUS EQUATIONS

There are several methods available for solving simultaneous equations, such as the pivotal reduction technique and the functional iteration techniques by Gauss, Seidel, Jacobi, Grant, Newton and Jordan. However, these are usually applicable only to linear or simple polynomial type of equations. The interested reader is referred to the books by Gerald (1970), Noble (1970), Cohen (1973) and Conte (1972).

The Newton-Raphson method for solving simultaneous equations uses an algorithm and a procedural technique which is similar to method we discussed in Section 1.1(A). The method is capable of solving two equations which can be non-linear, but suffers from the same limitation of having to take differentials.

As it is beyond the scope of this book to review existing techniques, we will first of all define the new method, before introducing the calculator technique of solving simultaneous equations using the same algorithm we employed to find the roots of a single equation. Subsequently, we will change the BASIC program given in Listing 2.1 in order to deal with more than one equation.

The procedure to be followed in order to solve two or more simultaneous equations is identical to the one used to find the roots of single equations. Again, we will use the same attenuation algorithm we employed when solving for the roots of single equations. In order to simplify explanation, we assume just two equations which are arranged in such a way as to assign x_0 to F_0 and x_1 to F_1. The method depends on the correct assignment of variables which we will refer to as the 'Equation Sequence' - we could quite easily have assigned x_0 to F_1 and x_1 to F_0, respectively, though convergency is not always guaranteed with any one sequence.

Similarly, a different sequence could give another genuine answer in some cases.

The next step is to assign the initial values for X_i and S_i. Then, we iterate once through F_0, as in the first line of Table 2.1, to obtain a better value for x_0 and using this we solve F_1 for a better value of x_1. This cyclic sequence is continued unitl convergence is achieved. Initially, we choose the sign of the attenuation algorithm to be -1 for all equations, but if this sign combination does not produce convergence, the sign of the last equation is reversed and the procedure repeated. Just as a different 'Equation Sequence' could give another answer, when solving complicated simultaneous equations, so can a different combination of signs.

A strict convention for changing the sign of the attenuation algorithm should be adopted so that no one combination is missed. We will adopt the binary sequence. Below, we show the set of sign combinations for two equations:

$$-1, -1\,;\ -1, +1\,;\ +1, -1\,;\ +1, +1$$

and shall refer to it as the 'Sign Combination'.

3.1 Using a calculator to solve simultaneous equations

We will illustrate the method with an example.

Problem:

Use the $\sinh^{-1}$ algorithm to find the coincidence points of the following equations, which are actually two quartics:

$$F_0 = x_0 + \sqrt{x_1} - 11 = 0$$

$$F_1 = \sqrt{x_0} + x_1 - 7 = 0$$

Solution:

Assign x_0 to F_0 and x_1 to F_1, respectively. Assume the initial pre-signs of H_0, and H_1 (see Section 2.2) to be +1 and that $x_0 = x_1 = 0$

and $S_0 = S_1 = -1$ as the initial parameters.

Then the following table is obtained in which you alternate the calculation in each line from the first to the second set of columns.

TABLE 3.1 Solving Simultaneous Equations by Calculator

$F_0 = x_0 + \sqrt{x_1} - 11$				$F_1 = \sqrt{x_0} + x_1 - 7$			
x_{0_n}	Sign of H_0 Pre-sign=+1	P_0	R_0	x_{1_n}	Sign of H_1 Pre-sign=+1	P_1	R_1
0	+	1	0	0	+	1	0
3.093	+	2	0	2.358	+	2	0
6.307	+	3	0	4.249	−	2	1
8.997	−	3	1	4.094	−	3	1
8.981	+	3	2	4.022	−	4	1
8.986	+	4	2	4.002	−	5	1
8.992	= ANSWER			4.000	= ANSWER		

The above method can be applied to any number of simultaneous equations as the procedure remains the same whatever their number or complexity. Remember that 'Sign Combination' as well as 'Equation Sequence' can be of importance in finding one or more solutions. When possible, linear equations should be sequenced so that the coefficients of the leading diagonal are greater than, or equal to, all other coefficients (see Section 3.3 for more details).

3.2 Computer method for solving simultaneous equations

The BASIC computer program shown in Listing 3.1 is capable of solving any number of simultaneous equations. It is based on the generalized ROOTS program, the only difference being that we had to employ arrays in order to provide the required suffixes for F, X and SIGN. As before, use the same line numbers as those given in the Listing in order to allow for future development.

```
  0  HOME:PRINT:PRINT "FOR ROOTS OF SINGLE EQU. USE
     LINE 10":PRINT
  1  PRINT "FOR SIMULTANEOUS EQUATIONS":
     PRINT "ENTER YOUR EQUATIONS AS FOLLOWS:":PRINT
  2  PRINT "10 F(0)=3 * X(0)– 7 * X(1) + 5:RETURN"
  3  PRINT "11 F(1)=2 * X(0) + 3 * X(1)–1:RETURN":PRINT
  7  PRINT "PRESS SPACE TO CONTINUE":PRINT "OR
     <RESET> TO ENTER EQUATION":GET T$:GOTO 100
100  DEF FNS(X) = LOG(ABS(X)+SQR(X*X+1))
110  HOME
140  INPUT "NR OF EQUATIONS? ";N:NN=N:N=N–1
150  DIM F(NN),X(NN),P(NN),S1(NN),H(NN),X1(NN),S(NN),
     S2(NN),X2(NN),R1(NN),A(NN),R(NN),A1(NN)
155  INPUT "DECIM ACCUR = ";D:INPUT "NR OF ITERATIONS
     ";M
170  FOR I=0 TO N:PRINT "X(";I;") = ";:INPUT "";X(I):
     X2(I)=X(I):NEXT I
180  FOR I=0 TO N:PRINT "R(";I;") = ";:INPUT "";R(I):
     R1(I)=R(I):NEXT I
190  FOR I=0 TO N:S1(I)=1:P(I)=0:A(I)=I:A1(I)=I:NEXT I
210  IF N=0 THEN INPUT "SIGN F(0) = ";S(0):GOSUB 320:END
220  PRINT
270  FOR I=0 TO N:PRINT "SIGN F(";I;") = ";:INPUT "";S(I):
     NEXT I:FOR I2=0 TO N:I=A(I2):X(I)=X2(I):R(I)=R1(I2):
     S1(I)=1:P(I)=0:NEXT I2:GOSUB 300:IF N=0 GOTO 270
275  GOTO 220
300  PRINT
320  PRINT:PRINT "ITER";TAB(10);"P";TAB(17);"R";TAB(28);
     "ROOT":PRINT
325  FOR K=1 TO M:L1=–1:FOR I2=0 TO N:I=A(I2):I1=A1(I2)
330  J=I+1:ON J GOSUB 10,11,12,13,14,15,16,17,18,19,20,
     21,22,23,24,25,26,27,28,29,30,31,32,33,34,35,36,
     37,38,39,40,41,42,43,44,45
333  IF ABS(X(I1))>1E6 OR ABS(F(I))1>E18 THEN PRINT
     "X OR F ABOVE LIMIT SET IN LINE 333":K=M:NEXT
     K:RETURN
335  H(I)=S(I)*FNS(F(I))*SGN(F(I)):S2(I)=SGN(H(I))
345  IF S2(I)*S1(I)>0 THEN P(I)=P(I)+1:R(I)=R(I)–1
400  R(I)=R(I)+1:X1(I1)=X(I1)+H(I)*2↑(P(I)/3-R(I)-1/3)
405  PRINT TAB(2);K;TAB(10);P(I);TAB(17);R(I);TAB(25);X1(I1)
410  IF ABS(X1(I1)–X(I1))<D THEN L1=L1+1
415  X(I1)=X1(I1):S1(I)=S2(I):NEXT I2:IF L1<>N GOTO 485
420  PRINT:PRINT "SOLUTION";TAB(22);" RESIDUAL":PRINT
425  FOR I3=0 TO N:I=A1(I3):I1=A(I3)
430  PRINT "X";I;"=";X1(I);TAB(21);"F";I;"=";
```

```
435  IF L1=N THEN J=I+1:ON J GOSUB 10,11,12,13,14,15,16,
     17,18,19,20,21,22,23,24,25,26,27,28,29,30,
     31,32,33,34,35,36,37,38,39,40,41,42,43,44,45
440  PRINT TAB (25);F(I)
445  NEXT I3
480  PRINT:K=M:NEXT K:RETURN
485  IF PEEK(-16384)>127 THEN POKE-16368,0:K=M:NEXT
     K:RETURN
490  IF N=0 THEN NEXT K:GOTO 500
495  PRINT:NEXT K
500  PRINT "NOT CONVERGING IN ";M;" ITERATIONS"
504  RETURN
```

Listing 3.1 Short SIM.ROOTS program for solving simultaneous equations.

The above program was developed on the Apple II microcomputer and line 485 is machine dependent. It simply awaits for any key of the keyboard to be pressed, in which case location -16384 will contain a value greater than 127. If that happens, the keyboard strobe is reset. This line adds greater flexibility to the program (to be discussed shortly) and if you are able to implement it on your machine you should do so. However, it is not essential to the functioning of the program and can be ignored by typing 485 REM, as the actual line number is required by the program. Finally, the command HOME, in lines 0 and 110, simply clears the screen and places the cursor on the top left-hand corner of the screen. Again, it is not essential to the functioning of the program and if you are unable to implement it on your machine, it can be left out. The above program has been implemented on the BBC microcomputer as shown in Appendix C. A very short but limited version of the program is given in Section B.2 of Appendix B.

On typing 'RUN' the following information will appear on the screen:

```
FOR ROOTS OF SINGLE EQU. USE LINE 10
FOR SIMULTANEOUS EQUATIONS
ENTER YOUR EQUATIONS AS FOLLOWS:

10 F(0)=3 * X(0) - 7 * X(1) + 5:RETURN
11 F(1)=2 * X(0) + 3 * X(1)- 1:RETURN

PRESS SPACE BAR TO CONTINUE
OR <RESET> TO ENTER EQUATION
```

To check the program type the two equations given above and RUN it with the initial parameters given below, noting that in general, for simultaneous equations, all Rs are set to zero:

```
NR OF EQUATIONS? 2
DECIM ACCUR = .000001
NR OF ITERATIONS 50
X(0) = 0
X(1) = 0
R(0) = 0
R(1) = 0
SIGN F(0) =-1
SIGN F(1) =-1
```

ITER	P	R	ROOT
1	0	1	-.917691763
1	1	0	1.76506344
2	0	2	-.320637798
2	1	1	.761480301
3	1	2	-.0522046622
3	2	1	.129550126
4	1	3	-.312105163
4	2	2	.456669889
-	-	-	———
-	-	-	———
24	15	9	-.347825727
24	17	7	.565217171

SOLUTION	RESIDUAL
X(0)= -0.347825727	F(0)=2.61888E-06
X(1)= 0.565217171	F(1)=5.96046E-08.

The program returns to that part of the initial input position which asks

SIGN F(0) =

SIGN F(1) = ,

where the signs of the function can be changed without having to re-enter the rest of the initial conditions. Now try the sign combination −1,+1 which should produce divergence. If you have implemented line 485, as soon as you detect a divergence in the iterations, you could press any key on the keyboard which would cause the computer to abandon the iterations with that particular sign combination and return the program to the input condition

SIGN F(0) =
SIGN F(1) =.

If line 485 is not implemented, you will have to wait for the completion of all iterations entered initially (in this case 50), before the program returns to the same input condition, or if an overflow is detected by line 333.

3.3 Sign combination and equation sequence subroutines

In the case of three or more simultaneous equations the cyclic order of solving for x_i from equation F_i is simply extended and solutions are found just as easily, provided we have found the correct SIGN combination. As the number of equations increase, the number of SIGN combinations escalates. If there are N equations, there are 2^N SIGN combinations for that particular equation sequence, hence the need for a simple optional computer SIGN SEARCH subroutine.

In exceptional cases the problem of convergence depends on the sequence of iterative cycling through the equations, or more appropriately the current assignment of an x_i variable for solution by an F_j equation. Those familiar with the classical Gauss-Seidel method for solving linear equations (Gerald, 1970) will recognise this standard well-conditioning sequence criteria (i.e. equations are rearranged so that the coefficients of the leading diagonal are greater than, or equal to, all other coefficients, as shown in Fig. 3.1). With the new algorithm, rearranging equations and using different sign combinations is sometimes essential, especially if searching for multiple roots or in order to converge on to a given solution in fewer iterations.

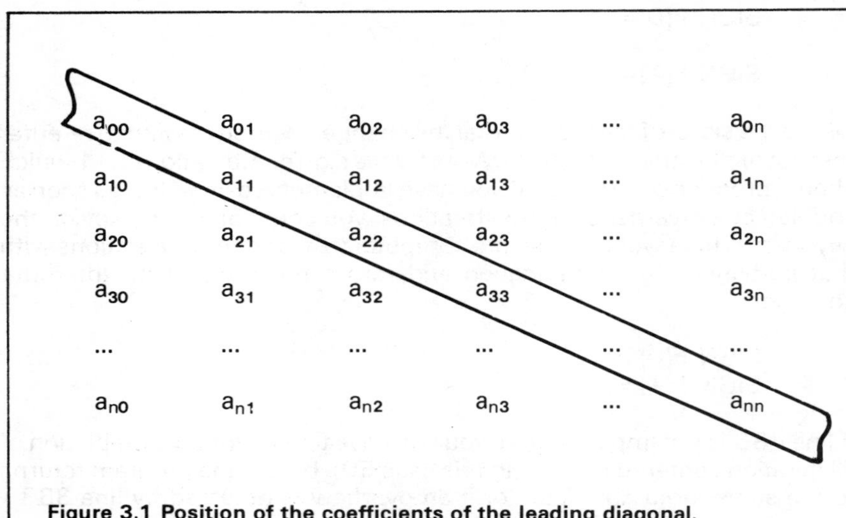

Figure 3.1 Position of the coefficients of the leading diagonal.

The program lines shown in Listing 3.2, when added to the program appearing in Listing 3.1 will not only provide the means for automatic choice of sign and equation sequence, but will also restore the Fine Search mode needed when searching for roots of single equations. Thus, the resulting program replaces all previous programs, as it can be used under all eventualities.

```
145  INPUT "CONSTANTS? (Y/N) ";T$:IF  LEFT$(T$,1)="Y"
     THEN
     INPUT "HOW MANY? ";G:DIM G(G):FOR I=0 TO G-1:
     PRINT "G(";I;")=";:INPUT "";G(I):NEXT I
160  INPUT "OUTPUT RESULTS TO PRINTER? (Y/N) ";TT$
200  K1=0:K2=0:IF N=0 THEN INPUT "FINE SEARCH? (Y/N)
     ";T$:IF LEFT$
     (T$,1)="Y" THEN INPUT "SEARCH INCREMENT? ";L:K1=1:
     K2=1
210  IF N=0 THEN INPUT "SIGN F(0) = ";S(0):IF K1=1 THEN
     PRINT:
     PRINT "F(X)";TAB(25);"X":PRINT:GOSUB 325:GOTO 270
215  IF N=0 THEN GOSUB 320:GOTO 270
220  INPUT "RE-ARRANGE EQUATIONS? (Y/N) ";T$:
     IF LEFT$(T$,1)="N" GOTO 250
```

```
240 PRINT "ENTER EQUATION SEQUENCE":PRINT:PRINT
    "ORIGINAL F()","NEW SEQUENCE":FOR I=0 TO N:PRINT
    I,:INPUT "";A(I):NEXT I
250 INPUT "SIGN SEARCH? (Y/N) ";T$:IF LEFT$(T$,1)="Y"
    THEN GOSUB 600:GOTO 240
300 PRINT:PRINT:PRINT "SIGN COMB.;
305 FOR I=0 TO N:PRINT TAB(3*I+12);S(I);:NEXT I:PRINT
310 PRINT:PRINT "EQU. SEQU.;
315 FOR I=0 TO N:PRINT TAB(3*I+12);A(I);:NEXT I:PRINT
340 IF K1=1 THEN GOSUB 505:IF K1=1 THEN GOSUB 10:
    GOTO 335
417 D$=CHR$(4):IF LEFT$(TT$,1)="Y" THEN PRINT D$;"PR
    #1"
450 PRINT:PRINT "SIGN COMB.;
455 FOR I=0 TO N:PRINT TAB(3*I+12);S(I);:NEXT I
460 PRINT:PRINT "EQU. SEQU.;
465 FOR I=0 TO N:PRINT TAB(3*I+12);A(I);:NEXT I:PRINT:PRINT
467 IF LEFT$(TT$,1)="Y" THEN PRINT D$;"PR#0"
470 PRINT "TO CONTINUE, TYPE ... CONT":PRINT:END
501 PRINT:PRINT "SIGN COMB.;
502 FOR I=0 TO N:PRINT TAB(3*I+12);S(I);:NEXT I
503 PRINT:PRINT "EQU. SEQU.;
504 FOR I=0 TO N:PRINT TAB(3*I+12);A(I);:NEXT I:PRINT:PRINT:
    PRINT "TO CONTINUE, TYPE ... CONT":END:RETURN
505 IF K2=1 THEN K2=0:S1(I)=S2(I):GOTO 515
510 PRINT F(I);TAB(25);X(I):IF S1(I)*S2(I)<=0 THEN K1=0:
    PRINT:PRINT "ITER";TAB(10);"P";TAB(17);"R";
    TAB(28);"ROOT":PRINT:RETURN
515 X(I)=X(I)+L:RETURN
600 T1=2↑NN-1:FOR T2=0 TO T1:T5=T2:FOR I2=0 TO N:
    I=A(I2):X(I)=X2(I):R(I)=R1(I2):S1(I)=1:P(I)=0:NEXT I2
610 FOR I=N TO 0 STEP-1:T3=T5/2:T4=INT(T3):
    IF T3-T4<0.001 THEN S(I)=(-1):GOTO 630
620 S(I)=1
630 T5=T4:NEXT I:GOSUB 300:NEXT T2
640 PRINT:PRINT "ALL SIGN COMBINATIONS EXHAUSTED":
    INPUT "RE-ARRANGE EQUATIONS? (Y/N) ";T$:
    IF LEFT$(T$,1)="N" THEN PRINT:PRINT "END OF
    PROGRAM":END
650 RETURN
```

Listing 3.2 Requisite additions to program in Listing 3.1 to form a general SIM.ROOTS program equipped with 'Sign Combination' and 'Equation Sequence' subroutines.

Line 145 allows the injection of constants into equations. The need for this will be discussed at the end of Section 5.1. Line 417 which is required in order to obtain a hard copy of the results of calculations, assumes that the printer peripheral-card is in slot 1 of the Apple II. At present, however, we ask you to respond with an N to both questions:

"OUTPUT RESULTS TO PRINTER? (Y/N)" and "CONSTANTS? (Y/N)".

For the sake of simplicity, both questions have been omitted from the printouts shown below until such time as we need their use.

We shall illustrate the use of both the 'Sign Combination' search mode and the 'Equation Sequence' facility by the example given below.

Problem:

Find the roots or coincidence points of the following three equations:

$$2X_1 + 3X_2 + X_3 - 1 = 0$$
$$3X_1 + X_2 + 2X_3 - 6 = 0$$
$$X_1 + 2X_2 + 3X_3 - 5 = 0$$

Solution:

Enter the equations into lines 10, 11 and 12 in the order they appear above. Supply the initial conditions as:

NR OF EQUATIONS? 3
DECIM ACCUR. = .000001
NR OF ITERATIONS 50
X(0) = 0
X(1) = 0
X(2) = 0
R(0) = 0
R(1) = 0
R(2) = 0
RE-ARRANGE EQUATIONS? (Y/N) N
SIGN SEARCH? (Y/N) Y

The Sign Combination subroutine will automatically start with SIGNs -1, -1, -1, the most commonly encountered combination for well conditioned equations. After a few iterations it will become obvious that the solutions will not converge, in which case we can go to the next sign combination by pressing any key on the computer's keyboard (provided line 485 was implemented), otherwise you will have to wait until the total number of iterations have finished. Again, with the new choice of SIGNs (−1, −1, +1) it will become obvious very quickly that no convergence is likely. Only when the third combination of SIGNs is reached (−1, +1, −1) will the answers oscillate around their correct values of 1.0, −1.0 and 2.0, but never reach sufficient accuracy. All other sign combinations cause the answers to diverge.

Once all SIGN combinations have been exhausted, the message

ALL SIGN COMBINATIONS EXHAUSTED
RE-ARRANGE EQUATIONS? (Y/N)

will appear on the screen. Responding with Y to the last question will cause the following message to appear on the screen:

```
ENTER EQUATION SEQUENCE
ORIGINAL F()                NEW SEQUENCE
0                           1
1                           0
2                           2
SIGN SEARCH? (Y/N) Y
```

with the user providing the numbers under the New Sequence column and the Y in response to the last question.

The program now restarts the SIGN combination from -1,-1,-1, but F(0) is used to solve for X(1), F(1) for X(0) and F(2) for X(2). In 19 iterations the correct answers are reached within the specified accuracy.

It should be pointed out here that some simultaneous equations can either have an infinite number of coincident roots, like the line of intersection of two planes, or no roots at all. In the former case, solving such sets of equations can usually produce convergency for every starting condition, while the latter case will result in divergency or in an unending oscillation.

Ill-conditioned simultaneous equations often take a long time to converge unless the correct equation sequence and function signs are chosen. The reason for this is because ill-conditioned equations have a small angle between their function curves, i.e. a small coefficient change, which results in a big displacement in the intersection point.

Whenever possible, the burden of equation sequencing and sign searching can be reduced by noting the following:

(a) Sequence the equations so that, in the case of linear equations, the Gauss-Seidel criteria tends to be satisfied,

(b) Avoid sequences in which an attempt is being made to solve x_i from F_j if F_j does not contain any x_i terms, and

(c) If, after the first few iteration steps, one or more of the values of x diverge with their corresponding Rs remaining constant, then go to the next sign combination. Line 333 has the effect of automatically returning the program to the next Sign Combination, if the newly calculated value of any X exceeds a chosen absolute limit (in this case, 1E6), or F(x)>1E18. Obviously, the values of these limits can be varied by the operator.

(d) If possible, reduce the number of equation by substitution.

In conclusion, the total number of sequence and sign-search permutations are limited because there are usually several such combinations yielding the same solutions in the case of linear equations or possibly different solutions for non-linear equations.

WARNING:

With linear equations ill-conditioning is signified by a very small determinant. Occasionally, when you try to solve such problems, the following can occur:

(i) A large number of iterations are needed, possibly over 100, depending on the requested accuracy.
(ii) The iterations oscillate forever around a coincidence point.
(iii) There is divergency for all sign and equation sequences, the less rapid usually indicating the correct combination.

The reason behind this behaviour is that basically the roots algorithm is a mathematical equivalent of a high gain servo system containing non-linear elements (in this case the $\sinh^{-1}$ function and the multiplier 2^q). In designing such systems for fast response (in this case mimimum iterations), one must have a 'wide band' capability together with a high servo 'loop gain'.

However, in the case of very small signal-to-noise ratio (in this case lines representing each equation being almost parallel near the coincidence points), such a system tends to oscillate.

One method of stopping oscillations is to change the slope difference of functions at the coincidence points by substituting $x_1 = x_0 x_1$. The resultant equations are easily solvable in very few steps. Remember to substitute back into the above expression in order to extract the correct value of x_1 from the ratio of the results. An alternative non-linearizing function of x_i is $\tan^{-1}(x_i)$.

Another way of stopping such oscillations is to cut down the 'loop gain' (in this case most easily identifiable by lines 345 and 400 in Listing 3.1) which, however, increases the response time of the system (by increasing the number of iterations). To achieve this, replace the two program lines by:

```
345 IF S2(I)*S1(I)>0 THEN P(I)=P(I)+GF:R(I)=R(I)-GF
400 R(I)=R(I)+GF:X1(I1)=X(I1)+H(I)*2↑(P(I)/3-R(I)-1/3)
```

and provide an input for the 'gain factor' GF at the end of line 140. The above problem can be investigated by considering the following two equations whose determinant is 0.015. Use GF=0.01.

$$x_0 + 0.9925\,x_1 - 0.3 = 0$$
$$x_0 + 1.0075\,x_1 - 0.5 = 0,$$

In the rare cases where, in spite of the above treatment, oscillations persist (assuming that there is a solution), replace the $\sinh^{-1}$ function in line 100 by X, thus removing the non-linear element.

For very slow iterations always set DECIM ACCUR a factor of 10 or 100 better than you need in running the SIM.ROOTS program.

In some cases, two or more quite simple non-linear simultaneous equations can have many valid solutions, possibly an infinite number. A simultaneous equation variation on the single equation of Section 2.3, which has the same roots, is:

$$F_0 = x_0 - x_1 = 0$$

$$F_1 = x_0^7 + 28x_1^4 - 480 = 0.$$

With the initial parameters $x_0 = x_1 = 0$, $R_i = 0$, GF=0.1 and $S_i = -1$,

SIM.ROOTS will give the first root $x_0 = x_i = 1.92288$. The next two roots are very close together and, as you can prove by experimenting, these two equations tend to be 'ill-conditioned' under such circumstances - but not insoluble.

In order to find the next set of roots, put $x_0 = x_1 = 0$, $R_i = 2$, $GF = 0.1$, $S_0 = -1$ and $S_1 = +1$ which will lead to $x_i = -2.4581$. Note that a slightly bigger R than R = 0 is needed to prevent initial oscillatory iterations (i.e. root jumping), which also leads to quicker and more reliable convergency on the next root.

In a manner identical to that for root searching in Sections 2.3 and 2.4, take at least one of the new starting values of x_i further along the axis, close to, yet, in the simultaneous case, as far away from the previous root (crossover) as you dare - otherwise, due to initial iteration oscillations, you might lock onto a previous root. In this case put $x_0 = -2.5$, x_1 = previous root $= -2.458$. Also change signs appropriately, and as before, set $R_i = 2$, $GF = 0.1$. With $S_i = +1$, the solution is $x_i = -2.5779$.

Alternatively, you can approach this root from a more negative initial $x_i = -4$. To arrive at the same solution, again put $R_i = 2$, $GF = 0.1$, and $S_i = +1$.

3.4 Solving complex and unrelated equations

The new algorithm can also be used to solve complex equations as well as multiple unrelated equations and, although we will use the computer program to illustrate the technique, what can be achieved by computer can also be done with a calculator. For functions of a complex variable, see Table A.3 of Appendix A.

Problem:

Find all the roots, whether real or complex, of the following equation:

$$F(x) = x^3 - 5x^2 + 24x - 20 = 0$$

Solution:

Use the SIM.ROOTS program and type the equation in line 10 as follows:

```
10 F(0) = X(0)↑3 -5*X(0)↑2 + 24*X(0) -20:RETURN
```

We could use the program to solve for the roots of this single equation which has only one real root at X(0)=1, by responding with 1 to the question NR OF EQUATIONS. Then, realizing that there are no more real roots, substitute

$x = x_1 + jx_2$ (where j is the square root of−1)

into the original equation and separate real and imaginary terms in order to obtain two equations which we could enter into lines 10 and 11 to solve for the complex roots.

Alternatively, we could solve for all three roots at once by entering the two extra equations into lines 11 and 12.

We shall choose the latter course in order to illustrate this method. Type the two extra equations as follows:

```
11 F(1) = X(1)↑3 - 5*X(1)↑2 + 24*X(1) - 3*X(1)*X(2)↑2 +
          5*X(2)↑2 - 20:RETURN
12 F(2) = X(2)↑2 - 3*X(1)↑2 + 10*X(1) - 24:RETURN
```

and choose the Sign Search mode. Convergency is achieved in 15 iterations with SIGNs -1, +1, -1 giving the three answers:

X(0)=1, X(1)=2 and X(2)=4

which indicates one real root at 1, and two complex roots at (2+j4) and (2-j4).

Another example, this time transcendental, of solving two simultaneous equations and one related independent equation at once, is given below by applying the method to a bridge-cable catenary problem.

Problem:

A bridge consists of a heavy cable of length s metres, suspended between two horizontal points 260m apart. The cable sag is 65m with its lowest point being d metres above ground level. Find the cable lengths.

Solution:

Figure 3.2 illustrates the bridge-cable catenary. In order to find the cable length s, we must first solve the two simultaneous equations: (See Appendix A for cosh and sinh definitions)

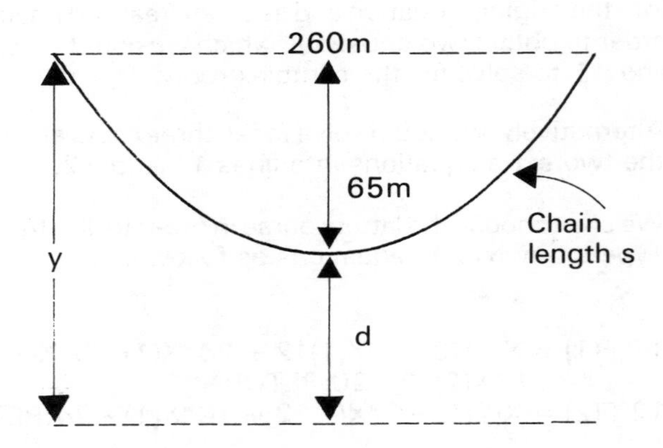

Figure 3.2 Bridge-cable catenary.

$$y = d \cosh\left(\frac{260}{2d}\right)$$

$$y = d + 65$$

for y and d, then substitute d into

$$s = 2\ d \sinh\left(\frac{260}{2d}\right)$$

to obtain the cable length s.

In order to solve these equations, we type them into the computer as follows (see Appendix A):

```
10 F(0)=X(0) - X(1)*(EXP(130/X(1)) + EXP(-130/X(1)))/2:
   RETURN
11 F(1)=X(0)-X(1)-65:RETURN
12 F(2)=X(2)-2*X(1)*(EXP(130/X(1))-EXP(-130/X(1)))/2:
   RETURN
```

where X(0), X(1) and X(2) have been substituted for y, d and s, respectively.

Providing the initial parameters for ACC. =.000001, all Xs = 100 (we choose a high initial value, such as 100 in order to avoid an overflow error which will result from attempting to evaluate EXP() with a large argument), all Rs = 0 and choosing the 'Sign Search' mode, we can proceed to solve for X(0), X(1) and X(2). When the SIGNs become -1, +1 -1, the following solutions are obtained:

X(0)=204.661625
X(1)=139.661625
X(2)=299.205689.

Thus, as y=X(0) and d=X(1), we see from the results that the sag in the cable(y-d) equals 65m, exactly the value given in Fig. 3.2. The chain length s=X(2)=299.205689m is precisely the answer to within the requested accuracy.

CHAPTER 4

SOLVING PRACTICAL PROBLEMS

In this Chapter we shall use examples taken from Physics, Management, Electronics and Mathematics in order to demonstrate the overall method of approach to solving practical problems.

Naturally, it will help if the reader has some understanding of the relevant subject areas as we shall mainly concentrate on the method of solution rather than the derivation of the equations.

The first four problems require us to find the roots of single equations, while the last two are examples requiring the solution of multiple equations. We shall use the SIM.ROOTS program in both cases, although it would have been just as easy to use the ROOTS program for the first four examples.

4.1 VARIATION OF VAPOUR PRESSURE WITH TEMPERATURE

The Kirchoff-Rankine-Dupré formula which gives the vapour pressure P (in millimetres of mercury) over the temperature range

$$15 \leqslant T \leqslant 270$$

can be expressed as:

$$\log_{10}P = 15.24431 - \frac{3623.932}{(273.16+T)} - 2.367233 \log_{10}(273.16+T),$$

where T is in degrees Centigrade.

To find the temperature T at which the vapour pressure P is, say, 2.25×10^{-3} mm Hg, substitute the value of P into the formula and write X(0) for T, to obtain the SIM.ROOTS format as:

```
10    F(0)=15.24431-3623.932/(273.16+X(0))-2.36723*
      (LOG(273.16+X(0))/LOG(10))-(LOG(2.25E-03)/LOG(10)):
      RETURN
```

Now type 'RUN' and provide the typical initial parameters as shown below. The full printout is included so that the reader can see all intermediate steps.

```
NR OF EQUATION? 1
DECIM ACCUR = .000001
NR OF ITERATIONS 50
X(0) = 0
R(0) = 0
FINE SEARCH? (Y/N) N
SIGN F(0) = -1
```

ITER	P	R	ROOT
1	1	0	.97839393
2	2	0	2.17445052
3	3	0	3.62405132
4	4	0	5.36119405
5	5	0	7.41202739
6	6	0	9.78544006
7	7	0	12.4597117
8	8	0	15.3663008
9	9	0	18.3753964
10	10	0	21.2932175
11	11	0	23.8844971
12	12	0	25.9275246
13	13	0	27.2904134
14	14	0	27.9957215
15	15	0	28.2282904
16	16	0	28.249567
17	16	1	28.2477859
18	17	1	28.2468516
19	18	1	28.2465402
20	19	1	28.2465113
21	19	2	28.2465138
22	20	2	28.2465151
23	21	2	28.2465155

```
SOLUTION                RESIDUAL

X(0)=28.2465155         F(0)=1.16415322E-09

SIGN COMB. -1

EQU. SEQU. 0

TO CONTINUE, TYPE ... CONT
```

So, the temperature is 28.246515 degrees Centigrade. The fact that the 'residual' (the difference between the given value of P and that which is calculated by substituting the value of T back into the original equation) is extremely small (1.164×10^{-9}), shows that the calculated value of T is very accurate.

The last line of the printout only appears if a solution is reached. Typing 'CONT' will cause the program to request SIGN F(0) = which allows you to look for another possible solution without having to re-enter the initial conditions.

4.2 TUNING SHARPNESS OF A RESONANT CIRCUIT

Let us calculate the tuning sharpness (otherwise known as the quality factor, Q), of an RLC electronic circuit (a circuit made up of resistive, inductive and capacitive components), as shown in Fig. 4.1(a). The voltage gain of such a circuit is given by:

$$|G| = \left|\frac{V_0}{V_i}\right| = \frac{R_C}{\{(R_L + R_c - \omega^2 LCR_C)^2 + \omega^2 (L + R_L R_c C)^2\}^{1/2}} \qquad (4.1)$$

where ω is the frequency of the applied alternating voltage V_i.

The variation of voltage gain with frequency is shown in Fig. 4.2(b).

Note that $|G|$ reaches a maximum value at ω_0, which is known as the resonant frequency of the circuit. The tuning sharpness of the circuit is measured, by definition, at a point where the gain is equal to $(1/\sqrt{2})$ of its maximum value. The width of the curve $(\omega_2 - \omega_1)$ at that point is the 'bandwidth' of the system.

We could write the voltage gain expression in complex form, and set the imaginary part equal to zero in order to solve for the resonant frequency, ω_0, of the circuit. This corresponds to equating the first denominator bracket in Equation (4.1) to zero and solving for wo to obtain

$$\omega_0 = \left\{ \frac{R_L + R_c}{LCR_c} \right\}^{1/2} \tag{4.2}$$

Substituting in Equation (4.2) the values L=1 H, C=1 μF, R_L =100 Ω and R_c =10 kΩ, we obtain:

ω_0 = 1000 $(1.01)^{1/2}$ rads/sec.

The remaining factors in the expression for G constitute the gain of the circuit at resonance, given by

$$G_0 = \frac{R_c}{\omega_0(L + R_L R_c C)} = \frac{5}{(1.01)^{1/2}} \tag{4.3}$$

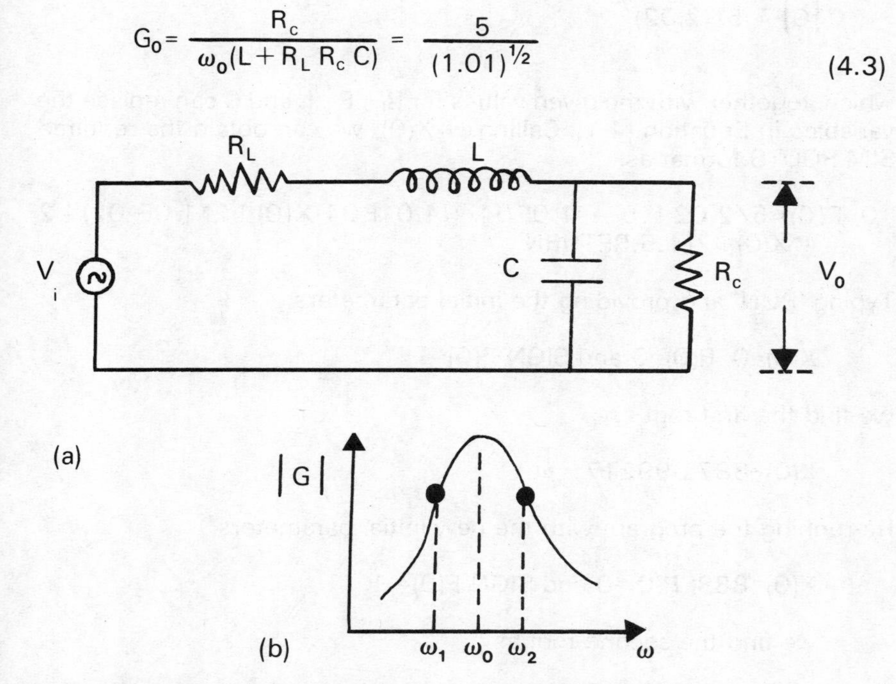

Figure 4.1 (a) RLC circuit, (b) variation of gain with frequency.

We can calculate the tuning sharpness (Q) of this circuit from

$$Q = \frac{\omega_0}{\text{Bandwidth}} = \frac{\omega_0}{(\omega_2 - \omega_1)} \tag{4.4}$$

where ω_1 and ω_2 are the two roots of the general voltage gain formula (Equation 4.1) when

$$|G| = \frac{G_0}{\sqrt{2}}$$

Substituting Equation (4.3) into the above expression, we obtain:

$$|G| = 5/(2.02)^{1/2},$$

which, together with the given values for R_L, R_c, L and C can replace the variables in Equation (4.1). Calling $\omega = X(0)$, we can obtain the required SIM.ROOTS format as:

```
10 F(0)=5/2.02↑.5 – 1.0E04/((1.01E04-X(0)↑2*1.0E–02)↑2
   +4*X(0)↑2)↑.5:RETURN
```

Typing 'RUN' and providing the initial parameters

X(0)=0, R(0)=0 and SIGN F(0)=1

we find the first root as

X(0)=887.699217 = ω_1.

Re-running the program with the new initial parameters

X(0)=888, R(0)=0 and SIGN F(0)=–1

we find the second root as

X(0)=1091.78299 = ω_2.

Substituting ω_1 and ω_2 into the formula for the tuning sharpness of the circuit (Equation 4.4), we can calculate Q as 4.924 which is far more accurate than would normally be required.

A full printout of both program runs is shown below.

NR OF EQUATIONS? 1	NR OF EQUATIONS? 1
DECIM ACCUR = .000001	
DECIM ACCUR = .000001	
NR OF ITERATIONS 50	NR OF ITERATIONS 50
X(0) = 0	(0) = 888
R(0) = 0	R(0) = 0
FINE SEARCH? (Y/N) N	FINE SEARCH? (Y/N) N
SIGN F(0) = 1	SIGN F(0) = -1

ITER	P	R	ROOT
1	1	0	1.65753924
2	2	0	3.7459066
3	3	0	6.37707827
4	4	0	9.692128
5	5	0	13.8687815
6	6	0	19.1309247
7	7	0	25.760564
8	8	0	34.1128554
9	9	0	44.6349584
10	10	0	57.8896263
11	11	0	74.5845909
12	12	0	95.6089015
13	13	0	122.077301
14	14	0	155.383216
15	15	0	197.259345
16	16	0	249.840533
17	17	0	315.71231
18	18	0	397.896997
19	19	0	499.63234
20	20	0	623.443859
21	21	0	767.34739
22	22	0	904.385271
23	22	1	887.177638
24	22	2	887.437854
—	—	—	—
31	28	3	887.699217

SOLUTION RESIDUAL

X (0)=887.699217 F(0)=0

SIGN COMB. 1
EQU. SEQU. 0

TO CONTINUE, TYPE ... CONT

ITER	P	R	ROOT
1	1	0	888.004701
2	2	0	888.010715
3	3	0	888.018443
4	4	0	888.02842
5	5	0	888.04138
6	6	0	888.058362
7	7	0	888.080815
8	8	0	888.110875
9	9	0	888.151734
10	10	0	888.20833
11	11	0	888.288567
12	12	0	888.405618
13	13	0	888.58244
14	14	0	888.861122
15	15	0	889.323376
16	16	0	890.138495
17	17	0	891.684185
18	18	0	894.877909
19	19	0	902.17823
20	20	0	920.94836
21	21	0	975.251476
22	22	0	1120.17559
23	22	1	1089.12317
24	22	2	1090.76684
—	—	–	———
32	28	4	1091.78299

SOLUTION RESIDUAL

X (0)=1091.78299 F(0)=–9.31323E–09

SIGN COMB.–1
EQU. SEQU. 0

TO CONTINUE, TYPE ... CONT

Although the number of iterations in the above and some later examples might be considered to be slightly larger than expected, one has to remember that there is usually a reason for this, namely:

(a) a very high accuracy might have been called for,
(b) a very high capture range might be required (as in the above example which stretches from 0 to 887),
(c) the equations might be simultaneous, and
(d) if simultaneous, the equations might not be well conditioned.

4.3 DIPOLE FIELD STRENGTH

An electric or magnetic dipole field strength due to 2 monopoles, $\pm m$, placed at points A and B (as shown in Fig.4.2(a)), and separated by a distance t, can be represented as the sum of two vectors having a resultant magnitude H. Fig.4.2(b) shows a sketch of the magnitude of the field strength H for constant values of H, assuming a unit value for m. The contours (curves joining points of equal field strength) are represented on an x, y distance axes centered at point A of Fig.4.2(a). The contours can be obtained by solving the equation:

$$(H/m)^2 = \frac{1}{(x^2 + y^2)} + \frac{1}{[(t-x)^2 + y^2]^2} + \frac{2\,[x\,(t-x) - y^2]}{\{(x^2 + y^2)\,[(t-x)^2 + y^2\,]\}^{3/2}}$$

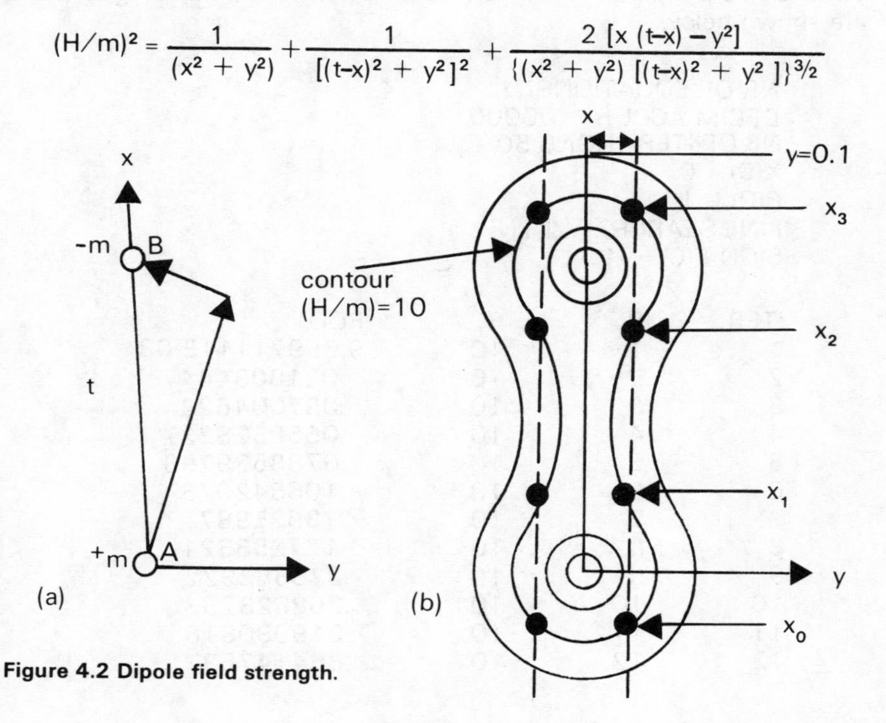

Figure 4.2 Dipole field strength.

for x at constant values (H/m), assuming a value for y.

As an example, we choose the contour (H/m)=10 (shown in Fig. 4.2(b)) and take the case when y=+0.1 with t=1.

Substituting these values into the equation for H/m and writing X(0) for x and Z for (H/m), we obtain:

```
10 Z=10:Y=.1:T=1
11 F(0)=Z↑2–1/(X(0)↑2+Y↑2)↑2–1/((T–X(0))↑2+Y↑2)↑2–
   2*(X(0)*(T–X(0))–Y↑2)/((X(0)↑2+Y↑2)*((T–X(0))↑2+Y↑2))↑1.5:
   RETURN
```

In Section 5.1 we will show that either (H/m) or y can be represented by 'constants' which are incremented, allowing contours to be plotted at preset intervals by solving the above equation repeatedly for x. However, returning to the present problem, the four roots (x_0, x_1, x_2 and x_3 shown in Fig.4.2(b) when y=+0.1) can be found quite easily by RUNning the SIM.ROOTS program four times. The full printout of these program runs are shown below.

```
NR OF EQUATIONS? 1
DECIM ACCUR = .000001
NR OF ITERATIONS 50
X(0) = 0
R(0) = 10
FINE SEARCH? (Y/N) N
SIGN F(0) = -1

ITER    P     R     ROOT
1       1     10    9.65971111E-03
2       2     10     .0218094447
3       3     10     .037004622
4       4     10     .0558328227
5       5     10     .0788539798
6       6     10     .106542078
7       7     10     .13925987
8       8     10     .177253321
9       9     10     .220582922
10      10    10     .268823753
11      11    10     .319990818
12      12    10     .364147537
```

13	12	11	.343413254
14	13	11	.333600438
15	13	12	.343121229
16	13	13	.340886978
17	13	14	.341052895
18	14	14	.341116404
19	15	14	.341125909
20	15	15	.341125385

SOLUTION RESIDUAL

X(0) = .341125385 F(0) = .3E-03

SIGN COMB. –1
EQU. SEQU. 0

TO CONTINUE, TYPE ... CONT

NR OF EQUATIONS? 1
DECIM ACCUR = .000001
NR OF ITERATIONS 50
X(0) = 0
R(0) = 10
FINE SEARCH? (Y/N) N
SIGN F(0) = 1

ITER	P	R	ROOT
1	0	11	–3.8334589E-03
2	1	11	–8.66150472E-03
3	2	11	–.0147364273
4	3	11	–.0223675043
5	4	11	–.0319266447
6	5	11	–.0438479409
7	6	11	–.0586163675
8	7	11	–.0767423036
9	8	11	–.0987241105
10	9	11	–.125006387
11	10	11	–.155933857
12	11	11	–.191666682
13	12	11	–.231937087
14	13	11	–.27517926
15	14	11	–.312746109
16	14	12	–.293682702
17	15	12	–.281231062

18	15	13	-.291391972
24	18	16	-.290775529

SOLUTION   RESIDUAL

X(0)=-.290775529   F(0)=2.2156E-06

SIGN COMB. 1
EQU. SEQU. 0

TO CONTINUE, TYPE ... CONT

NR OF EQUATIONS? 1
DECIM ACCUR = .000001
NR OF ITERATIONS 50
X (0) = .35
R (0) = 10
FINE SEARCH? (Y/N) N
SIGN F(0) = 1

ITER	P	R	ROOT
1	1	10	.352430963
2	2	10	.355768615
3	3	10	.360337935
4	4	10	.366564736
5	5	10	.375000001
6	6	10	.386347848
7	7	10	.401496397
8	8	10	.421549444
9	9	10	.447853341
10	10	10	.482003454
11	11	10	.525779206
12	12	10	.580806243
13	13	10	.646840295
14	14	10	.701367632
15	14	11	.657860979
16	14	12	.661158201
17	14	13	.658006202
18	14	14	.658725558
19	15	14	.658891023
20	15	15	.658882147
21	16	15	.658877173
22	17	15	.658875289
23	18	15	.658875007

```
SOLUTION                              RESIDUAL

X(0)=.658875007                       F(0)=.2E-04

SIGN COMB.    1
EQU. SEQU.    0

TO CONTINUE, TYPE ... CONT




NR OF EQUATIONS? 1
DECIM ACCUR = .000001
NR OF ITERATIONS 50
X (0) = .7
R (0) = 10
FINE SEARCH? (Y/N) N
SIGN F(0) = -1
```

ITER	P	R	ROOT
1	1	10	.70427402
2	2	10	.709821919
3	3	10	.717059337
4	4	10	.726553048
5	5	10	.739083557
6	6	10	.755739746
7	7	10	.778069102
8	8	10	.808333206
9	9	10	.849984159
10	10	10	.908649184
11	11	10	.994113538
12	12	10	1.11671186
13	13	10	1.24318605
14	14	10	1.34582696
15	14	11	1.30080661
16	15	11	1.26140981
17	15	12	1.28972022
18	16	12	1.29846697
19	16	13	1.28702303
20	16	14	1.29152782
21	16	15	1.29069351
—	—	—	—
25	19	16	1.29077553

```
SOLUTION                     RESIDUAL
X(0)=1.29077553              F(0)=.6E-05

SIGN COMB.   -1
EQU. SEQU.   0

TO CONTINUE, TYPE ... CONT
```

Thus, the four roots are:

$x = -0.291$, $x = 0.341$, $x = 0.659$ and $x = 1.291$.

Note that as the contours are completely symmetrical about points A and B of Fig.4.2(b), we now know the exact position of all eight points marked on the figure. The left four (which correspond to $y=-1$) are the mirror image of the four points we have just found.

WARNING:

Although it is extremely practical to enter constants into the computer as we did in the previous example (line 10), you must not use variable names which are part of the program itself. A list of all variables used in the various programs discussed in this text appear in Appendix D.

Also note that the above facility can only be used with single equations, hence the RETURN at the end of line 11 in the previous example. For simultaneous equations, any such constants must appear as the first statement of a relevant line, with the second statement on the same line being the equation itself, as the program expects to find a subroutine on each successive line from line number 10 to 45. Simultaneous equations which are longer than 256 characters each can be solved either by incorporating a GOSUB statement in lines 10, 11, 12, etc., in order to divert program execution to, say, lines 10000, 11000, 12000, etc., or by the use of the DIFF.ROOTS version of the program as discussed in Chapter 8.

4.4 DISCOUNTED CASH FLOW

The pattern of cash flow associated with a certain project can be represented by the diagram of Fig. 4.3. The construction of buildings and the installation of plant and machinery takes three years to complete (represented in the diagram by Years−2 to 0).

The project starts operating from Year 1, but because of additions to the capacity of the plant in that year, a net cash outflow is expected. The project starts to earn profits in Year 2, but a cyclic pattern is expected because of the price of raw material needed for the production of goods. This cyclic pattern gives sequences of positive and negative cash flows which might result in more that one Rate of Return that could be earned on the project. Our object is to find the project's Rate of Return and ensure that no other Rate of Return exists.

The first step in solving this problem is to relate the cash flows arising from the project to a common base year, namely, Year 0. Cash flows arising in years prior to Year 0 must be adjusted by an interest or earnings factor which must be added to the cost of the investment, in order to bring it to the correct value for the base year. Cash flows arising in years subsequent to Year 0 must be brought back to their value at the base year by means of discount factors that will reduce the value of the future receipts to their present values at the base year. The procedure is known as the Discounted Cash Flow (DCF) calculation.

If the outflows of Years 5 and 6 were the result of planned additions to capacity of the project, then the usual mehtod of discounting these outflows to previous years' inflows could be employed in order to achieve a continuous inflow of cash after Year 0 (apart from the final year when the tax payable on the last year's profits could exceed the residual values). This method would have simplified the problem by ensuring only one DCF solution.

Since the present value (PV) of an amount \$C received after n years at X% interest rate, is given by

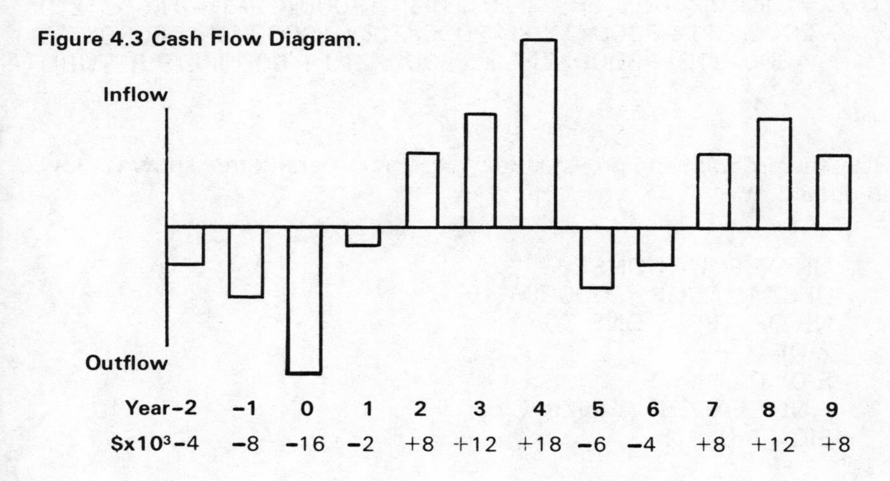

Figure 4.3 Cash Flow Diagram.

$$PV = \frac{C}{(1 + X/100)^n}$$

and the DCF solution we are seeking is the value of X when the the sum of all present value (PV) cash flows is equal to zero, then we can write the Net Present Value (NPV) as:

$$NPV = -\frac{4000}{(1+X/100)^{-2}} - \frac{8000}{(1+X/100)^{-1}} - \frac{1600}{(1+X/100)^{0}} - \frac{2000}{(1+X/100)^{1}}$$

$$+\frac{8000}{(1+X/100)^{2}} + \frac{12000}{(1+X/100)^{3}} + \frac{1800}{(1+X/100)^{4}} - \frac{6000}{(1+X/100)^{5}}$$

$$-\frac{4000}{(1+X/100)^{6}} + \frac{18000}{(1+X/100)^{7}} + \frac{12000}{(1+X/100)^{8}} + \frac{8000}{(1+X/100)^{9}} = 0$$

which we can solve for all positive values of X (only positive values of X have meaning here), using the SIM.ROOTS program.

Writing X(0) for X and then replacing the denominator of each element of the above equation by Z, where Z = 1 + X(0)/100, we write the equation into line 10 in the following manner:

```
10  Z=1 + X(0)/100:F(0)=-4000/Z↑(-2) - 8000/Z↑(-1) - 16000/Z↑0
    -2000/Z↑1 + 8000/Z↑2 + 12000/Z↑3 + 18000/Z↑4-6000/Z↑5
    - 4000/Z↑6 + 8000/Z↑7 + 12000/Z↑8 + 8000/Z↑9:RETURN
```

RUN the program and provide the typical initial parameters shown below, to obtain:

```
NR OF EQUATIONS? 1
DECIM ACCUR = .000001
NR OF ITERATIONS 50
X(0)=0
R(0)=0
FINE SEARCH? (Y/N) N
SIGN F(0) = 1
```

ITER	P	R	ROOT
1	1	0	10.858999
2	2	0	21.4399349
3	2	1	15.1766877
4	3	1	8.10022836
5	3	2	11.8788399
6	4	2	15.4938959
7	4	3	13.2402462
8	5	3	10.7823848
9	5	4	12.1131862
10	6	4	13.4134597
11	6	5	12.6219379
—	—	—	—
31	15	16	12.3583566

SOLUTION RESIDUAL

X(0)=12.3583566 F(0)=1.445E-04

SIGN COMB. 1
EQU. SEQU. 0

TO CONTINUE, TYPE ... CONT

Thus, a percentage rate of 12.36 will equate the present value of cash flows from the project with the value of the cash investment, and is the Rate of Return that would be earned on the funds invested in the project.

Employing the 'Fine Search' mode with an increment of 0.1 and starting from X(0)=0 with SIGN=+1 confirms that there is no other root between 0 and 12.358%. Repeating the exercise, but with X(0)=12.4 and SIGN=–1, reveals no further roots.

Traditional methods for finding the DCF solution of such a problem depend on a trial and error technique. A guess is made of the expected Rate of Return and the cash flows of all the years of the project are discounted to the base year using the guessed value. If the sum of the present value of all the cash flows is positive, the Rate of Return is increased by, say, 5 until that sum becomes negative. From that point on, several techniques can be used to find the value of the Rate of Return for which the sum of the present value of cash flows nears zero.

One such method is the interpolation technique. Thus, traditional methods for finding the DCF Rate of Return will lock into one of the solutions, depending on which one happens to be within the range of the trial values employed by the user. The user is uncertain of the existance of other solutions which must be found if they exist.

4.5 DIRECT CURRENT NETWORK

Figure 4.4 shows a direct current (D.C.) network with three loop currents (i_0, i_1 and i_2) driven by a 10 V source. We wish to find the values of the currents i_0, i_1 and i_2.

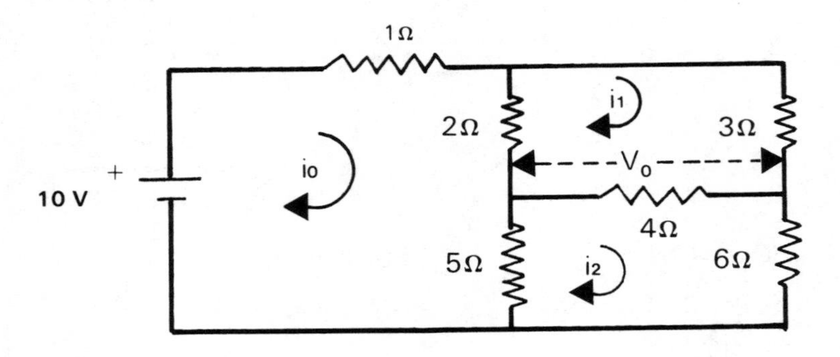

Figure. 4.4 Direct current network.

Using Kirchhoff's second theorem, we can write the potentials encountered in each loop as:

$$8 i_0 - 2 i_1 - 5 i_2 - 10 = 0$$
$$-2 i_0 + 9 i_1 - 4 i_2 = 0$$
$$-5 i_0 - 4 i_1 + 15 i_2 = 0$$

Adopting the SIM.ROOTS format for expressing such equations, remembering that the currents i_0 i_1 and i_2 must be expressed as X(0), X(1) and X(2), we write:

```
10 F(0) =  8*X(0) - 2*X(1) - 5*X(2) -10:RETURN
11 F(1) = -2*X(0) + 9*X(1) - 4*X(2):RETURN
12 F(2) = -5*X(0) - 4*X(1) + 15*X(2):RETURN
```

and use the program to solve for X(0), X(1) and X(2).

To find the voltage Vo, across the 4 Ω resistor, we would then substitute the values of i_0, i_1 and i_2 into the expression

$$V_0 = 4\, i_1 - 4\, i_2.$$

With the new method it is possible to solve the three simultaneous equations as well as the related equation concurrently, by formatting the latter as:

```
13 F(3)= X(3)– 4*X(1) + 4*X(2):RETURN
```

where X(3) represents the value of V_o.

The output of the program is shown below. Note that since the three simultaneous equations satisfy the Gauss-Seidel well- conditioned criteria of the diagonal coefficients being larger than all others, we find no need for either the 'Equation Sequence' or the 'Sign Search' mode. Solutions are rapidly reached with all SIGNs set to–1, as shown below.

```
NR OF EQUATIONS? 4
DECIM ACCUR = .000001
NR OF ITERATIONS 50
X(0) = 0
X(1) = 0
X(2) = 0
X(3) = 0
R(0) = 0
R(1) = 0
R(2) = 0
R(3) = 0
RE-ARRANGE EQUATION? (Y/N) N

SIGN SEARCH? (Y/N) Y
SIGN COMB.–1–1–1–1
EQU. SEQU. 0 1 2 3
```

ITER	P	R	ROOT
1	1	0	2.99822295
1	1	0	2.49119539
1	1	0	3.91065848
1	0	1	–.967277422
2	2	0	6.84287476
2	2	0	5.80601208
2	1	1	3.39584039
2	0	2	–.360698438
3	2	1	4.65304765
3	2	1	3.24003884
3	2	1	1.26469773
3	1	2	.341417537

4	3	1	1.98381705
4	3	1	.306387159
4	3	1	−.921657145
4	2	2	1.04211751
5	4	1	−1.00098799
5	4	1	−2.52394976
5	4	1	−1.9853169
5	2	3	.746215038
6	4	2	−.0865069355
6	4	2	−.836236752
6	4	2	−9.46658E-03
6	3	3	.328022513
7	5	2	1.73431505
7	5	2	1.10983227
7	5	2	2.056038
7	4	3	−.202408576
8	6	2	3.99730958
8	6	2	3.11705948
8	6	2	3.05331771
8	4	4	−.147062782
9	6	3	3.81426633
9	6	3	2.0049271
9	6	3	1.61558846
9	5	4	.0581559856
10	7	3	2.40026057
10	7	3	.698560464
10	7	3	.145230531
10	6	4	.357966207
11	8	3	.727586063
11	8	3	−.658277316
11	8	3	−.484091134
11	6	5	.266736698
12	8	4	.862718542
12	8	4	.111396679
12	8	4	.518015509
12	7	5	.0923826405

```
 ===          ===          ===          ===

 39           27           12           2.0272564
 39           27           12           .851788222
 39           27           12           .902895781
 39           29           10           -.204430525

 SOLUTION                  RESIDUAL

 X(0)=2.0272564            F(0)=-4.1388E-06
 X(1)=.851788222           F(1)=-1.925E-06
 X(2)=.902895781           F(2)=1.81538E-06
 X(3)=-.204430525          F(3)=-2.8964E-07

 SIGN COMB. - 1 - 1 - 1 - 1
 EQU. SEQU. 0 1 2 3

 TO CONTINUE, TYPE ... CONT
```

4.6 SIMULTANEOUS COMPLEX EQUATIONS

The solution of electromagnetic induction problems, where currents, fields and mutual coupling coefficients are complex quantities, requires the manipulation of complex simultaneous equations. We will assume here that the analysis of a geophysical induction method, using finite element techniques, gave the following set of equations:

$$A_{00} Z_0 + A_{01} Z_1 + A_{02} Z_2 - B_0 = 0$$
$$A_{10} Z_0 + A_{11} Z_1 + A_{12} Z_2 - B_1 = 0$$
$$A_{20} Z_0 + A_{21} Z_1 + A_{22} Z_2 - B_2 = 0.$$

All the components of the above equations may be replaced by complex quantities such as:

$$Z_0 = x_0 + j\, x_3$$
$$Z_1 = x_1 + j\, x_4$$
$$Z_2 = x_2 + j\, x_5, \text{ where } j = \sqrt{-1},$$

with analogous complex expressions for the constants A_{ij} and B_i.

Thus, we obtain the following set of equations, with typical A_{ij} coefficients, as a result of substituting these complex quantities into the three simultaneous equations:

$$(2+j3)(x_0+jx_3) + (4+j1)(x_1+jx_4) + (1)(x_2+jx_5) - (13+j17) = 0$$
$$(6+j2)(x_0+jx_3) + (3+j3)(x_1+jx_4) - (1)(x_2+jx_5) - (15+j20) = 0$$
$$(3+j3)(x_0+jx_3) + (3+j1)(x_1+jx_4) + (3)(x_2+jx_5) - (11+j20) = 0.$$

Resolving the above expressions into real and imaginary parts, we obtain six simultaneous equations which can then be typed into the computer in the usual SIM.ROOTS format as follows:

```
10 F(0)=6*X(0) + 3*X(1)-X(2) - 2*X(3) -3*X(4) - 15:RETURN
11 F(1)=2*X(0) + 4*X(1) + X(2) - 3*X(3) - X(4) -13:RETURN
12 F(2)=3*X(0) + 2*X(1) + 3*X(2) - 3*X(3) - X(4) - 11:RETURN
13 F(3)=2*X(0) + 3*X(1) + 6*X(3) + 3*X(4) - X(5) - 20:RETURN
14 F(4)=3*X(0) + X(1) + 2*X(3) + 4*X(4) + X(5) - 17:RETURN
15 F(5)=3*X(0) + X(1) + 3*X(3) + 2*X(4) + 3*X(5) - 20:RETURN
```

As the equations happened to be well conditioned, the initial parameters can be set to their usual values, i.e., all Xs and Rs are set to zero and SIGNs are set to -1. Note that by choosing the 'Sign Search' mode all SIGNs are automatically set to -1 which reduces the effort of entering initial parameters.

The linearity of the above equations was chosen intentionally in order to demonstrate the procedure. It must be stressed, however, that the method can deal equally well with non-linear equations.

For functions of complex variables, see Table A.3 in Appendix A.

```
NR OF EQUATIONS? 6
DECIM ACCUR = .000001
NR OF ITERATIONS 50
X(0) = 0
X(1) = 0
X(2) = 0
X(3) = 0
X(4) = 0
X(5) = 0
R(0) = 0
R(1) = 0
R(2) = 0
R(3) = 0
```

R(4) = 0
R(5) = 0
RE-ARRANGE EQUATIONS? (Y/N) N
SIGN SEARCH? (Y/N) Y

SIGN COMB. -1 -1 -1 -1 -1 -1
EQU. SEQU. 0 1 2 3 4 5

ITER	P	R	ROOT
1	1	0	3.40230665
1	1	0	2.52340272
1	0	1	-.855006381
1	1	0	2.42820858
1	0	1	-.221212744
1	1	0	.415424714
2	1	1	1.92110914
2	2	0	5.85168027
2	0	2	-.484270798
2	1	1	.730812747
2	0	2	.212509534
2	2	0	3.20482627
3	2	1	-.105991661
3	2	1	4.15933608
3	1	2	.171654564
3	1	2	1.35029812
3	1	2	.851505555
3	3	0	4.35575601
4	2	2	.790601208
4	3	1	3.78607894
4	2	2	.926314813
4	2	2	1.57114246
4	1	3	.843767285
4	3	1	2.42835584
5	3	2	1.74530848
5	4	1	2.90184244
5	3	2	1.61381355
5	2	3	1.36421602
5	1	4	.864057439
5	4	1	1.38476816

6	4	2	2.6146488
6	5	1	2.30257679
6	3	3	1.39576153
6	3	3	1.12389518
6	1	5	.856770181
6	4	2	1.67514167
—	—	—	—
—	—	—	—
—	—	—	—
—	—	—	—
—	—	—	—
—	—	—	—
35	24	11	2.00000137
35	24	11	3.00006084
35	26	9	1.00002208
35	24	11	1.00002004
35	25	10	1.00001576
35	25	10	1.99991883

SOLUTIONS	RESIDUAL
X(0) =2.00000137	F(0)=8.1316E−05
X(1) =3.00006084	F(1)=1.923E−04
X(2) =1.00002208	F(2)=1.1615E−04
X(3) =1.00002004	F(3)=4.3394E−04
X(4) =1.00001576	F(4)=8.6904E−05
X(5) =1.99991883	F(5)= −8.69E−05

SIGN COMB. -1 -1 -1 -1 -1 -1
EQU. SEQU. 0 1 2 3 4 5

TO CONTINUE, TYPE ...CONT

CHAPTER 5

TURNING POINTS

In this chapter we shall investigate the turning points (maxima or minima) of functions representing either simple curves or surfaces. Traditional methods for locating minima of functions of one variable, employing such methods as the Univariate and Fibonacci searches are well covered elsewhere (see Cohen, 1973).

5.1 FINDING THE TURNING POINTS OF FUNCTIONS

In this section we will show how any of the programs developed so far can be used under 'Fine Search' mode to plot the slope of an explicit function F (a function of the form F(x)=0), as well as evaluate the actual value of x at the turning point. The method accommodates multiple, close-multiple and slope singularty (vertical slope) turning points along any axis. Using the proposed scheme, we tend to trade off, on occasions, 2 to 3 decimal points of accuracy out of 9, for the sake of simplicity and flexibility. However, should it be possible to differentiate the function manually, then the computer can be relegated to solving accurately the resultant equations, rather than taking differentials.

Consider a multi-dimensional surface defined by the implicit function

$F[x_0, x_1, x_2, ...]=0.$

Taking differentials:

$$dF=\frac{\partial F}{\partial x_0}\,dx_0+\frac{\partial F}{\partial x_1}\,dx_1+\frac{\partial F}{\partial x_2}\,dx_2+...=0.$$

If only two variables or dimensions are involved and we are seeking a turning point parallel to the x_1-axis at A (Fig. 5.1), then

$$dF = \frac{\partial F}{\partial x_0} dx_0 + \frac{\partial F}{\partial x_1} dx_1 = 0,$$

or

$$\frac{dx_1}{dx_0} = - \frac{\partial F}{\partial x_0} \Big/ \frac{\partial F}{\partial x_1}$$

which equals zero at the turning point i.e.,

$$\frac{\partial F}{\partial x_0} = 0 .$$

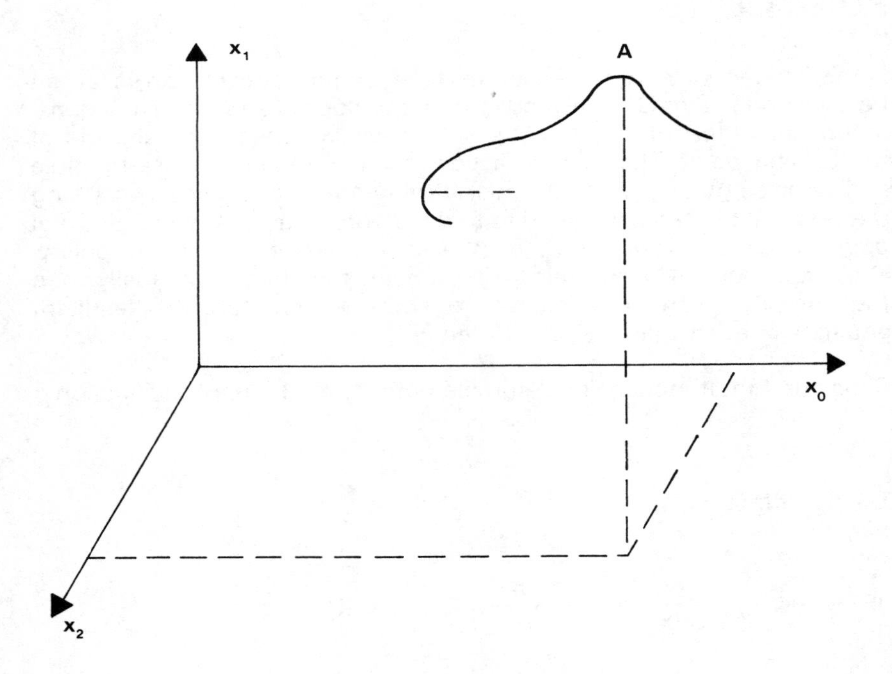

Figure 5.1 Turning points.

This must be solved simultaneously with

$F[x_0, x_1]=0$

in order to exactly determine the $[x_0,x_1]$ co-ordinates of the turning point at A.

Likewise, for more than two dimension (variable) surfaces, the following typical equations must be solved simultaneously:

$\partial F/\partial x_0 = 0$

$\partial F/\partial x_1 = 0$

$F[x_0, x_1, x_2, \ldots] = 0,$

with equation $\partial F/\partial x_1 = 0$ missing if the turning point sought is in the x_1-axis direction as shown in Fig. 5.1. If such a turning point exists, then the above is true for any x_i.

Replacing the partial differentials by difference equations, we obtain:

$F_0 = F[(x_0+G), x_1, x_2, \ldots] - F[(x_0-G), x_1, x_2, \ldots]/(2G) = 0$

$F_2 = F[x_0, x_1, (x_2+G), \ldots] - F[x_0, x_1, (x_2-G), \ldots]/(2G) = 0$

$F_n = F[x_0, x_1, x_2, \ldots] = 0,$

where G is a small increment of optional value, ideally as small as possible, but often limited in value by the computer round off errors. We find that G = .001 or G = .0001 as satisfactory without generally being critical.

In order to illustrate the formatting procedure, we will find both turning points of Fig. 2.1.

Problem:

Compute both turning points of the function

$F(x) = x^7 + 28x^4 - 480$

which lie between roots X=-2.57 and X=1.92 (see Fig. 2.1).

Solution:

Although we can format the above equation for use with the ROOTS program, we shall choose the SIM.ROOTS format as it paves the way to multi-dimensional turning points. Thus, the formatting is as follows:

```
10 G= .0001 : U=X(0)+G:W=X(0)-G:F(0)=((U↑7+28*U↑4-480)-
(W↑7+28*W↑4-480)) / (2*G):RETURN
```

Note that should the equation be very long, we could insert it into the computer using several lines in which case only the last line should be terminated with the RETURN statement as shown below:

```
10 G=.0001:U=X(0)+G:W=X(0)-G
11 Y1=U↑7+28*U↑4-480
12 Y2=W↑7+28*W↑4-480
13 F(0)=(Y1-Y2)/(2*G):RETURN
```

However, for multi-dimensional turning points, we must start each equation on successive line numbers, in which case the schemes discussed in section 4.3 (under the sub-heading 'Warning'), must be employed.

First, we will find the turning point which lies between roots x=-2.57 and x=-2.45. RUN the program and supply the initial parameters as follows:

```
NR OF EQUATIONS? 1
DECIM ACCUR = .000001
NR OF ITERATIONS 50
X(0) = -3
R(0) = 5
FINE SEARCH? (Y/N) Y
SEARCH INCREMENT? .1
SIGN F(0) = 1
```

F(X)	X
1432.18905	-2.9
914.607644	-2.8
507.447123	-2.7
193.897486	-2.6
-41.0145521	-2.5

ITER	P	R	ROOT
1	0	5	-2.60931297
2	0	6	-2.53385215
3	1	6	-2.46958461
4	1	7	-2.51089415
5	2	7	-2.54661304
6	2	8	-2.52308499
7	3	8	-2.50672977
8	3	9	-2.51914344
9	4	9	-2.52378642
10	4	10	-2.51825352
11	4	11	-2.52014051
12	4	12	-2.51984732
13	5	12	-2.51983999
14	5	13	-2.51984146
15	6	13	-2.51984215

SOLUTION	RESIDUAL
X(0)=-2.51984215	F(0)=0

SIGN COMB. 1
EQU. SEQU. 0

TO CONTINUE, TYPE ... CONT

The first five numbers appearing under the first column (when the 'Fine Search' mode is under operation), as well as the 'residual', give the slope of the equation at the corresponding values of x. The point at which this slope changes sign indicates the position of either a maximum or a minimum. In this case, as we iterate from left to right on the x-axis, the change of sign indicates a maximum.

We can now proceed to find the minimum turning point of the equation with a starting value for X(0) at say 1, and a 'Fine Search' increment of − 0.1. If we retain the value of G as 0.0001 with the same accuracy, we don't seem to get an accurate enough result. This is due to the fact that the slope of the equation near the minimum turning point is very shallow (RUN the program with 'Fine Search' and verify it). As a result of this, we need to increase both the value of G and the accuracy by, say 10, the choice being purely arbitrary. This was implemented in finding the minimum turning point of the equation as shown below.

```
NR OF EQUATIONS? 1
DECIM ACCUR = .000001
NR OF ITERATIONS 50
X(0) = 1
R(0) = 5
FINE SEARCH? (Y/N) Y
SEARCH INCREMENT?-.1
SIGN F(0) =-1

F(X)                                X

85.3681974                          .9
59.1791012                          .8
39.2396385                          .7
24.5186319                          .6
14.1094159                          .5
7.19672791                          .4
3.02911433                          .3
.896495767                          .2
.112035545                          .0999999997
0                                   -2.91038305E-10
-.111992005                         -.1

ITER        P           R           ROOT

1           0           5           -.0972280207
2           1           5           -.0940163798
3           2           5           -.0903564045
4           3           5           -.0862620087
5           4           5           -.081773967
6           5           5           -.0769524469
7           6           5           -.0718879067
8           7           5           -.0666871267
9           8           5           -.0614572856
10          9           5           -.0562991435
```

11	10	5	-.0512950156
12	11	5	-.0465329818
13	12	5	-.0420619861
14	13	5	-.0379001863
15	14	5	-.034038721
16	15	5	-.0305366286
17	16	5	-.0273580292
18	17	5	-.0244371627
19	18	5	-.0218507326
20	19	5	-.0194618903
21	20	5	-.0173333558
22	21	5	-.0155297272
23	22	5	-.0137332025
24	23	5	-.0122113107
25	24	5	-.0110182549
26	25	5	-9.63617184E-03
27	26	5	-8.43226673E-03
28	27	5	-8.43226673E-03

```
SOLUTION                      RESIDUAL
 X(0)=-8.43226673E-03         F(0)=0

SIGN COMB.-1
EQU. SEQU. 0

TO CONTINUE, TYPE ... CONT
```

Analytically, by differentiating the equation, we can get at any turning point,

$7\ x^6 + \ 112\ x^3 = 0,$

yielding a maximum at x = −2.51984210 and a minimum at 0, thus verifying the computer result.

It must be re-emphasised that should functions be easy to differentiate manually, then the computer should be used to solve the resulting equation which saves time. In the above case we chose such an equation only in order to compare the results of the two methods.

Note: The choice of variable names G, U and W, appearing in line 10, is purely arbitrary. However, such variables must not be given names which correspond to variables already used in the program. For this reason, we list in Appendix D all variable names which appear in all the various versions of the program.

In order to facilitate changes to the value of G or increase the number of constants, we shall use the facility provided by line 145 of the program which causes the question:

"CONSTANTS? (Y/N)"

to appear on the screen. Now type Y and then provide the number of constants you intend to use.

Note that when using this facility with the above program, the assignment of constant G in line 10 should be omitted, and G(0) substituted in place of G in all relevant equation lines.

5.2 MULTI-DIMENSIONAL TURNING POINTS

In order to illustrate the multi-variable format of turning points, given in Section 5.1, consider an example of a two-variable implicit function whose turning point we wish to locate.
Given the equation

$$F(x_0, x_1) = x_0^5 + 2\,x_0\,x_1 + x_1^5 = 0$$

we type into the computer the following two lines:

```
10 F(0)=((X(0)+G(0))↑5 + 2*(X(0)+G(0))*X(1) + X(1)↑5 -
   (X(0)-G(0))↑5 - 2*(X(0)-G(0))*X(1) - X(1)↑5) /
   (2*G(0)):RETURN
11 F(1)=X(0)↑5 + 2*X(0)*X(1) + X(1) ↑5:RETURN
```

and provide the usual initial parameters as:

```
NR OF EQUATIONS? 2
CONSTANTS? (Y/N) Y
HOW MANY? 1
G(0)=.0001
DECIM ACCUR = .0000001
NR OF ITERATIONS 50
OUTPUT RESULTS TO PRINTER? (Y/N) N
```

```
X(0) = -1
X(1) = -1
R(0) = 5
R(1) = 5
RE-ARRANGE EQUATIONS? (Y/N) N
SIGN SEARCH? (Y/N) N
SIGN F(0) = 1
SIGN F(1) = -1

SIGN COMB. 1 -1
EQU. SEQU. 0 1

ITER        P           R           ROOT

1           1           5           -.943173552
1           0           6           -1.00173023

2           2           5           -.88716453
2           1           6           -1.00512763

3           3           5           -.840454249
3           2           6           -1.00989003

4           4           5           -.811784964
4           3           6           -1.01570628

5           5           5           -.800799476
5           4           6           -1.02241697

6           6           5           -.799671694
6           5           6           -1.02989073

-           -           -           __________
-           -           -           __________

18          16          7           -.80814744
18          16          7           -1.06635709

SOLUTION                      RESIDUAL

X(0)=-.80814744        F(0)=0
X(1)=-1.06635709       F(1)=2.13876774E-08

SIGN COMB. 1 -1
EQU. SEQU. 0 1
TO CONTINUE, TYPE ... CONT
```

In order to find out whether the turning point at the above coordinate points is a maximum or a minimum, substitute X (1) =−1.06635709 into equation of line 10 and then solve for the root of that equation alone using the fine search mode. Obviously, the value of X(0) will be the same as that found previously, but the change of sign of the remainder, while the fine search mode is under operation, will reveal the type of turning point as discussed earlier. Naturally, there could be several more roots, i.e. turning points, in the above two equations.

5.3 TURNING POINTS OF SURFACES

In this section we will show that the method for finding the turning points of functions with three or more variables is identical to the one employed to find the turning point of the two-variable function discussed previously.

Consider the elliptic paraboloid given by:

$$F = (r\cos\theta - 3)^2/9 + (r\sin\theta - 2)^2/4 - 2(z-1) = 0$$

which typically has cylindrical coordinates given by:

$x = r\cos\theta = x_0\cos x_1$
$y = r\sin\theta = x_0\sin x_1$
$z = x_2$.

The minimum of this surface is at x=3, y=2 and z=1, as shown in Fig. 5.2 (given for comparison purposes only).
Thus, in order to locate the position of the minimum in terms of implicit variables r and θ, we follow the concepts set out in Section 5.1 and write the following three simultaneous equations:

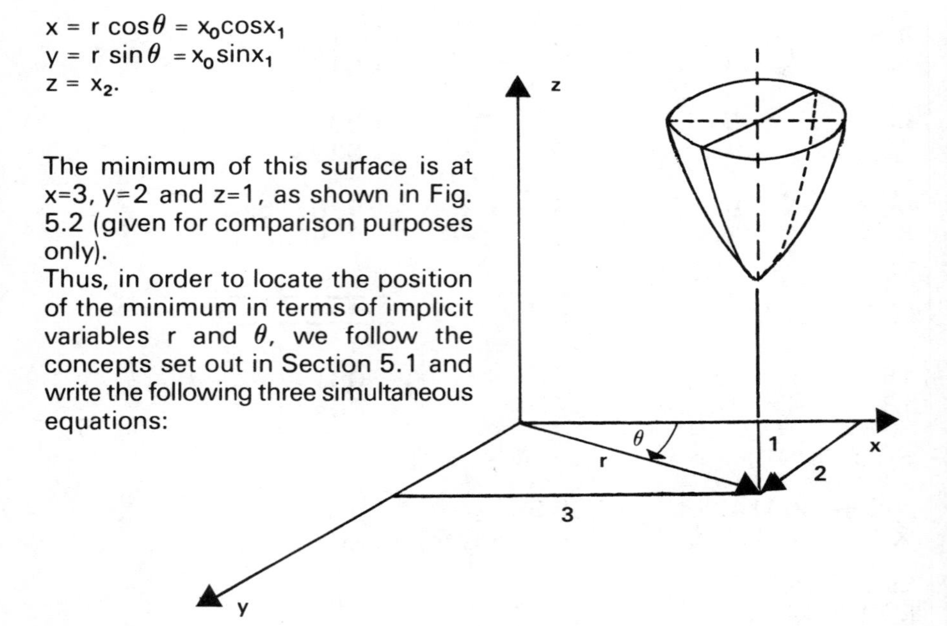

Figure 5.2 Elliptic paraboloid surface.

$$\frac{\partial F}{\partial F} = 0 = \frac{\partial F}{\partial x_0},$$

$$\frac{\partial F}{\partial \theta} = 0 = \frac{\partial F}{\partial x_1},$$

together with the original equation of the surface:

$F(r,\theta,z) = F(x_0, x_1, x_2) = 0.$

We leave out the $\partial F/\partial z$ equation, as described earlier, because the desired turning point is in the z direction. Thus,

$$\frac{\partial F(r,\theta)}{\partial r} = \frac{F[(r+G), \theta] - F[(r-G), \theta]}{2G}$$

and

$$\frac{\partial F(r,\theta)}{\partial \theta} = \frac{F[r, (\theta+G)] - F[r, (\theta-G)]}{2G}$$

As the equations tend to be rather long, we will adopt the formatting scheme proposed in Section 4.3 (under 'Warning').

Thus, with an axis increment G(0) and omitting cancelling terms, LOAD the SIM.ROOTS program and type:

```
10 GOSUB 10000:RETURN
11 GOSUB 11000:RETURN
12 GOSUB 12000:RETURN

10000 U1=X(0)+G(0):U2=X(0)–G(0)
10010 F1=(U1 * COS(X(1)) – 3)↑2/9 + (U1 * SIN(X(1)) – 2)↑2/4
10020 F2=(U2 * COS(X(1)) – 3)↑2/9 + (U2 * SIN(X(1)) – 2)↑2/4
10030 F(0)=(F1 – F2) / G(0):RETURN
11000 U3=X(1)+G(0):U4=X(1)–G(0)
11010 F3=(X(0) * COS(U3) – 3)↑2/9 + (X(0) * SIN(U3) – 2)↑2/4
11020 F4=(X(0) * COS(U4) – 3)↑2/9 + (X(0) * SIN(U4) – 2)↑2/4
11030 F(1)=(F3 – F4) / G(0):RETURN
12000 F(2)=(X(0) * COS (X(1)) – 3)↑2/9 +
        (X(0) * SIN(X(1)) – 2)↑2/4 – 2*(X(2) – 1):RETURN
```

Note that the first equation starts on line 10000, the second on line 11000 and the third on line 12000. Now RUN the program and provide the initial parameters shown below, to obtain:

```
NR OF EQUATIONS? 3
CONSTANTS? (Y/N) Y
HOW MANY? 1
G(0)=.0001
DECIM ACCUR = .0000001
NR OF ITERATIONS 100
OUTPUT RESULTS TO PRINTER? (Y/N) N
X(0) = 0
X(1) = 0
X(2) = 0
R(0) = 0
R(1) = 0
R(2) = 0
RE-ARRANGE EQUATIONS? (Y/N) N
SIGN SEARCH? (Y/N) N
SIGN F(0) =-1
SIGN F(1) =-1
SIGN F(2) = 1


SIGN COMB.-1-1 1
EQU. SEQU. 0  1  2
```

ITER	P	R	ROOT
1	1	0	1.09860907
1	1	0	152850874
1	1	0	1.87173292
2	2	0	2.17039268
2	1	1	.648827896
2	1	1	1.28964462
3	3	0	3.41395904
3	2	1	.387053721
3	2	1	1.0150162
—	—	—	—
—	—	—	—
—	—	—	—
21	16	5	3.6055513
21	13	8	.588002601
21	16	5	1

```
SOLUTION                    RESIDUAL

X(0)=3.6055513              F(0)=9.34190659E-09
X(1)=.588002601             F(1)=-2.4524896E-08
X(2)=1                      F(2)=4.59701721E-17

SIGN COMB.-1-1 1
EQU. SEQU. 0  1  2

TO CONTINUE, TYPE ... CONT
```

Note that the cartesian coordinates of the turning point can be found by substituting the above solutions into the x, y, and z expressions to obtain:

$x = r \cos\theta = X(0) \cos X(1) = 3,$
$y = r \sin\theta = X(0) \cos X(1) = 2,$
$z = X(2) = 1.$

However, establishing whether the turning point is a maximum or a minimum can sometimes be difficult because as with, say, a hyperboloid (see Fig. 5.3), a minimum as seen along the x-axis can be a maximum if viewed along the y-axis. With a slightly rotated hyperboloid, possibly both the x- and y-axis curves could appear as straight lines with no turning points. Because of this, you might find that the number of iterations required for convergency will differ as the function is viewed along different directions.

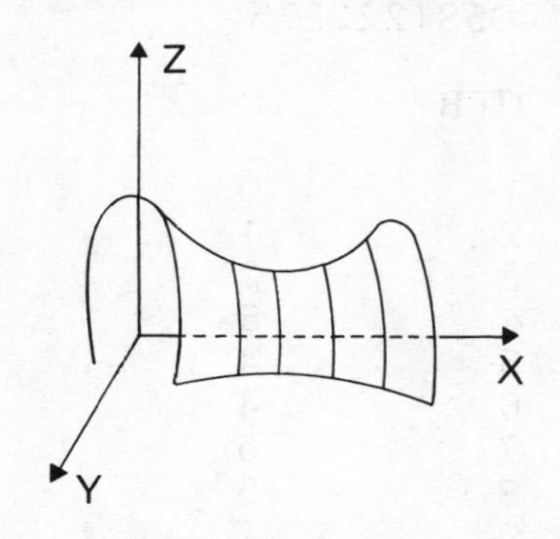

Figure 5.3 A hyperboloid

Returning to our problem, in order to find out whether the turning point of the function is a maximum or a minimum relative to r (=X(0)), substitute θ (=X(1)) into the first equation (lines 10 and 10000 – 10030), as shown below:

```
10 X(1)=.588002601:GOSUB 10000:RETURN
```

and with this first equation in lines 10000 to 10030 exactly as before, RUN the program in a single equation, Fine Search mode, as in Section 5.1, and supply the following initial parameters:

```
NR OF EQUATIONS? 1
CONSTANTS? (Y/N) Y
HOW MANY? 1
G(0)=.0001
DECIM ACCUR = .0000001
NR OF ITERATIONS 100
OUTPUT RESULTS TO PRINTER? (Y/N) Y
X(0) = 3
R(0) = 0
FINE SEARCH? (Y/N) Y
SEARCH INCREMENT? .1
SIGN F(0) = - 1

F(X)                         X

-.311108451                  3.1
-.249569275                  3.2
-.188031336                  3.3
-.12649296                   3.4
-.0649545654                 3.5
-3.41616266E-03              3.6
.0581222275                  3.7

ITER      P        R         ROOT

1         0        0         3.65389429
2         1        0         3.62414916
3         2        0         3.60972984
4         3        0         3.60564797
5         4        0         3.60552897
6         4        1         3.6055427
7         5        1         3.60554935
8         6        1         3.60555124
9         7        1         3.60555129


SOLUTION                     RESIDUAL

X(0)=3.60555129              F(0)=6.6297792E-09

SIGN COMB.-1
EQU. SEQU. 0

TO CONTINUE, TYPE ... CONT
```

The above solution clearly shows that as X(0)=r increases, the slope of the function goes from negative to positive, indicating a minimum as seen along the r direction.

Similarly, in searching for the type of turning point with respect to θ (= X(1)), we have to substitute r (=X(0)) into the second equation and solve as before. However, since the single equation mode of the program requires an equation to be typed in line 10 and the solution appears as X(0), we must change the subscripts of lines 11000-11030 from X(0) to X(1), X(1) to X(0) and F(1) to F(0). Thus, retyping equation 10 as:

```
10 X(1)=3.6055513:GOSUB 11000:RETURN
```

and using a Fine Search increment of 0.01, we can find that the slope of the function goes from negative to positive which indicates a minimum as viewed along an increasing θ-axis. Note that we chose a small Fine Search increment because angle θ is also small (.588).

Note: If, while under Fine Search mode, the value of the function approaches zero and then increases again without a change of sign, a point of inflection is indicated.

Finally, it must be pointed out that an alternative method of determining whether the function has a maximum or a minimum turning point, is by evaluating the second derivative and observing its sign. This can be done by extending the method we have adopted in finding the first derivative by differentiating once more as indicated below:

$$\frac{\partial^2 F(r,\theta)}{\partial r^2} = \frac{F[(r+2G),\theta] - 2F(r,\theta) + F[(r-2G),\theta]}{4G^2}$$

$$\frac{\partial^2 F(r,\theta)}{\partial \theta^2} = \frac{F[r, (\theta+2G)] - 2F(r,\theta) + F[r, (\theta-2G)]}{4G^2}$$

5.4 TURNING POINTS WITH CONSTRAINTS

Functions represented by F(x,y,z, ...) can have turning points when x, y, z, ... are not independent, but are subject to one or more constraints of the form:

$F_0 (x,y,z, ...) = 0.$

A function with three variables and one constraint can be expressed as:

$\delta F = 0$, or

$$\frac{\partial F}{\partial x}\delta x + \frac{\partial F}{\partial y}\delta y + \frac{\partial F}{\partial z}\delta z = 0.$$

The infinitesimal variations δx, δy and δz are not independent, but are connected by the relation:

$\delta F_0 = 0$, or

$$\frac{\partial F_0}{\partial x}\delta x + \frac{\partial F_0}{\partial y}\delta y + \frac{\partial F_0}{\partial z}\delta z = 0.$$

The Lagrangian method of undetermined multipliers allows us to form one equation from the above two relationships, namely,

$$\left(\frac{\partial F}{\partial x} - u\frac{\partial F_0}{\partial x}\right)\delta x + \left(\frac{\partial F}{\partial y} - u\frac{\partial F_0}{\partial y}\right)\delta y + \left(\frac{\partial F}{\partial z} - u\frac{\partial F_0}{\partial z}\right)\delta z = 0,$$

where u is an arbitrary multiplier which can be chosen to make each bracketed coefficient vanish, since there is no particular relationship between, say, δx and δy. (Note that with two constraint equations, there must be two such Lagrangian multipliers, and so on).

Thus, we can write:

$F_0 = 0,$

$$F_1 = \frac{\partial F}{\partial x} - u\frac{\partial F_0}{\partial x} = 0,$$

$$F_2 = \frac{\partial F}{\partial y} - u\frac{\partial F_0}{\partial y} = 0,$$

$$F_3 = \frac{\partial F}{\partial z} - u\frac{\partial F_0}{\partial z} = 0,$$

which constitutes four simultaneous equations in x, y, z and u, although the value of u need only be a check on convergency.

In order to illustrate the method, consider a rectangular box with edges $x=x_0, y=x_1$ and $z=x_2$ which has a volume of 27 cm. We are required to find the dimesions of the box which will give the least surface area. Thus, we must minimize the function:

$$F = 2 (x_0x_1 + x_0x_2 + x_1x_2),$$

subject to the following constraint on the volume:

$$F_0 = x_0x_1x_2 - 27 = 0.$$

Earlier in this chapter we showed how to obtain the differential coefficients. We proceed in identical manner and format the three Lagrange multiplier equations, as well as the constraint volume formula for F_0, in the usual manner. Denoting u by X(3), LOAD the SIM.ROOTS program and type:

```
10 GOSUB 10000:RETURN
11 GOSUB 11000:RETURN
12 GOSUB 12000:RETURN
13 GOSUB 13000:RETURN

10000 F(0)=X(0)*X(1)*X(2)-27:RETURN
11000 U1=X(0)+G(0)-U2=X(0)-G(0)
11010 F1=2 * ((U1*X(1) + U1*X(2) + X(1)*X(2)) -
       (U2*X(1) + U2*X(2) + X(1)*X(2)))
11020 F2= X(3) * ((U1*X(1)*X(2)-27)- (U2*X(1)*X(2)-27))
11030 F(1)=(F1-F2)/G(0):RETURN
12000 U3=X(1)+G(0):U4=X(1)-G(0)
12010 F3=2 * ((X(0)*U3 + X(0)*X(2) + U3*X(2))-
       (X(0)*(U4 + X(0)*X(2) + U4*X(2)))
12020 F4  X(3) * ((X(0)*U3*X(2)-27)-(X(0)*U4*X(2)-27))
12030 F(2)=(F3-F4)/G(0):RETURN
13000 U5=X(2)+G(0):U6=X(2)-G(0)
13010 F5=2 * ((X(0)*X(1) + X(0)*U5 + X(1)*U5)-
       (X(0)*X(1) + X(0)*U6 + X(1)*U6))
13020 F6=X(3) * ((X(0)*X1)*U5-27)-(X(0)*X(1)*U6-27))
13030 F(3)=(F5-F6)/G(0):RETURN
```

Note that lines 11010, 12010 and 13010 correspond to the first term of each of the Lagrange multiplier equations, respectively, while lines 11020, 12020 and 13020 correspond to the second term.

Now RUN the program and provide the initial parameters shown below. Note the need for equation sequence and sign search for convergency (see Section 3.3).

```
NR OF EQUATIONS? 4
CONSTANTS? (Y/N) Y
HOW MANY? 1
G(0)=.001
DECIM ACCUR = .0001
NR OF ITERATIONS 100
OUTPUT RESULTS TO PRINTER? (Y/N) N
X(0) = 1
X(1) = 1
X(2) = 1
X(3) = 1
R(0) = 0
R(1) = 0
R(2) = 0
R(3) = 0
RE-ARRANGE EQUATIONS? (Y/N) Y
ENTER EQUATION SEQUENCE
ORIGINAL F()              NEW SEQUENCE
0                         0
1                         2
2                         1
3                         3
SIGN SEARCH? (Y/N) N
SIGN F(0) =-1
SIGN F(1) = 1
SIGN F(2) =-1
SIGN F(3) = 1


SIGN COMB.-1 1 -1 1
EQU. SEQU. 0 2 1 3


ITER        P         R          ROOT

1           0                    4.95161334
1           0         1          -.320145943
1           1         0          2.9264476
1           1         0          4.77082198
```

```
2          2          0          10.1775027
2          0          2          .897967127
2          1          1          1.43858401
2          1          1          2.54474667

3          3          0          15.4523203
3          1          2          2.02611334
3          2          1          .8942151
3          2          1          -.722620594

4          3          1          14.7548957
4          1          3          1.389137
4          2          2          1.86643884
4          2          2          .927429097

-          -          -          ----------
-          -          -          ----------
-          -          -          ----------
-          -          -          ----------

66         48         19         3.00015625
66         47         20         3.0000419
66         48         19         2.99984964
66         47         20         1.33330223


SOLUTION                         RESIDUAL

X(0)=3.00015625                  F(0)=4.3E-04
X(1)=3.0000419                   F(1)=9.94E-04
X(2)=2.99984964                  F(2)=5.48E-04
X(3)=1.33330223                  F(3)=-2.446E-04

SIGN COMB.-1 1 -1 1
EQU. SEQU. 0 2 1 3

TO CONTINUE, TYPE ... CONT
```

Thus, the box sides are $x_0=x_1=x_2=3$ cm, whilst the Lagrangian multiplier is $x_3=1.3333$.

We should stress again that the problem used to illustrate the Lagrangian technique would be easier to solve by direct differentiation. The method shown is for problems which are difficult to differentiate.

CHAPTER 6

DATA FITTING

Historically, data fitting and interpolation has been the pursuit of mathematicians since before Newton's time. Tables of functions had to be interpolated for maximum precision, frequently by fitting a polynomial to adjacent points in a table. These were termed 'interpolation' or 'collocation' polynomials and were evaluated by the same means as polynomials fitted to 'exact' scientific data points, in order to reduce experimental results to an easily manipulated algebraic relationship, and for the purpose of interpolating intermediate values of a function.

When experimental data is given at equally spaced values of the independent variable t, there are several methods available for fitting a polynomial to the specific data values $t_i, f(t_i)$, based on finite differences. Such methods can employ Forward, Backward or Central Difference operators, to name but a few, in evaluating the coefficients of the fitting polynomial (Bartlett, 1974). When experimental data is given at unequal values of t, a popular scheme for evaluating the coefficients of the polynomial is the Langrangian method (Cohen, 1973 and Gerald, 1970).

In the above cases, the resultant polynomial passes through each data point. But Least Squares methods allow us to statistically fit the best straight line, parabola or other polynomial to a set of 'inexact' data points. Generally, a formula exists for straight line and quadratic Least Squares fitting coefficients.

All these methods are well covered in other texts (see Conte, 1972, Noble, 1972 and Watson, 1974).

In this chapter we will confine ourselves to fitting polynomials, including 'spline cubics' to unequally spaced data points.

The resultant simultaneous equations to be solved in terms of unknown coefficients are usually, but not always, linear.

6.1 INTERPOLATING POLYNOMIALS

Given a set of data points it is possible to determine a simple continuous function which passes through these points. One way of describing such a function is by an nth degree polynomial, as given below:

$$f(z) = a_0 + a_1 z + a_2 z^2 + a_3 z^3 + \dots + a_{n-1} z^{n-1} + a_n z^n, \qquad (6.1)$$

where a_0, a_1, ... a_n are the coefficients of the polynomial which must be evaluated.

In general, we need $(n+1)$ data points in order to fit an nth degree polynomial through them, unless we substitute the identity

$$z = t - (a_{n+1}/na_n) \qquad (6.2)$$

into Equation (6.1), which will result in an nth degree 'normal-form' polynomial in t, but with the (n–1)th term of the polynomial missing. This has advantages because we only require n data pairs and the higher degree polynomial fits the data points $(t_i, f(t_i))$ more accurately.

As an example, let us assume that we are required to fit a polynomial to the following four data points:

t_i	$f(t_i)$
0.1	3.56482944
0.7	3.96665924
1.6	2.06200548
2.2	-0.233496575

Experimental data are not usually as accurate as the above. In fact, we have generated these from the function:

$$f(t) = 4 \sin(t+1),$$

in order to have some means of comparison. Incidentally, this function has a root at $t=(\pi-1)$, as shown in Fig, 6.1.

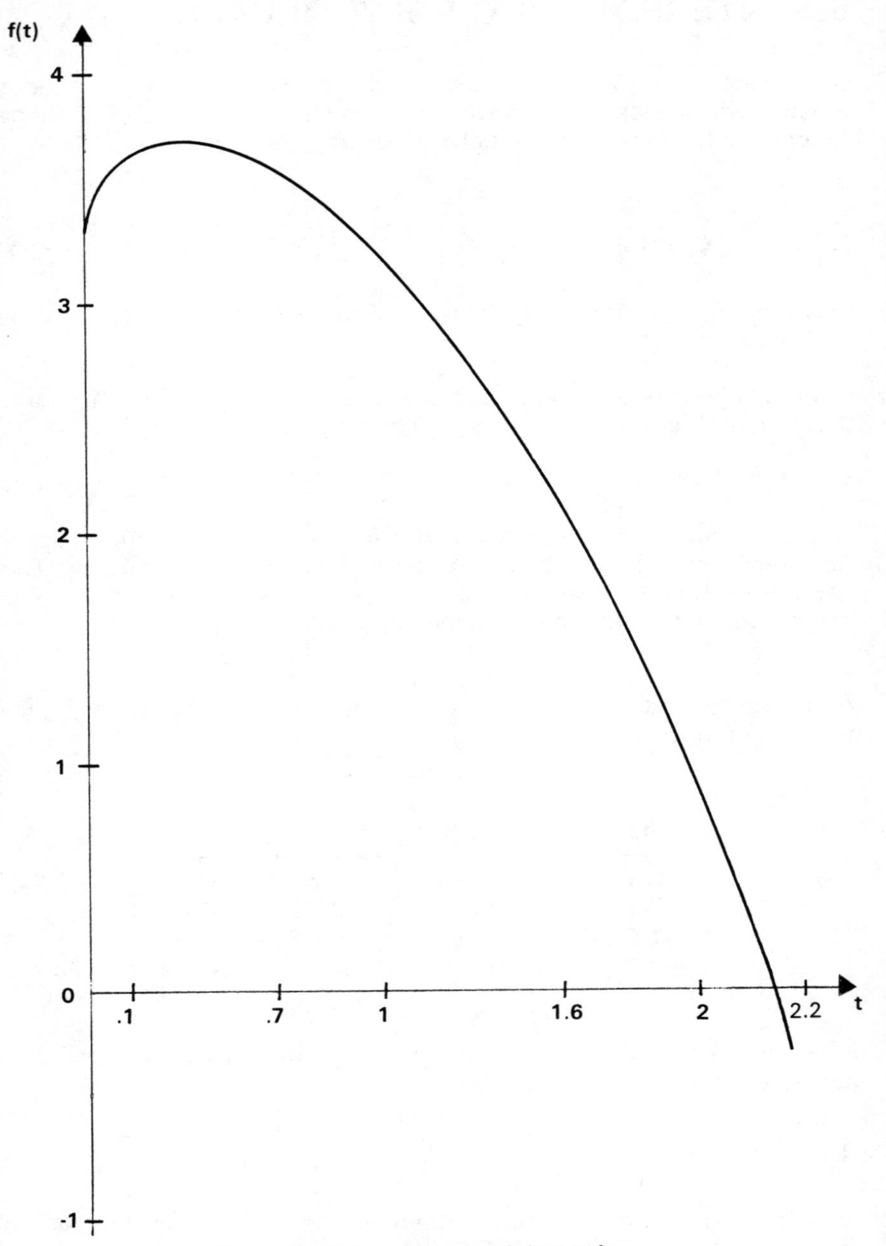

Figure 6.1 Unequally spaced data points.

Since in this case we have four data points, we write a fourth degree polynomial which is a normal-form quartic, as given below:

$$f(t) = b_0 + b_1 t + b_2 t^2 + b_3 t^4 \tag{6.3}$$

with b_i being the transformed a_i coefficients of Equation (6.1), after substituting into it Equation (6.2). Thus, we can write

$$F_i = b_0 + b_1 t_i + b_2 t_i^2 + b_3 t_i^4 - f(t_i) = 0,$$

and after substituting the coordinates of each data point, we obtain four simultaneous equations given by:

$$F_0 = b_0 + b_1 (0.1) + b_2 (0.1)^2 + b_3 (0.1)^4 - 3.56482944 = 0$$
$$F_1 = b_0 + b_1 (0.7) + b_2 (0.7)^2 + b_3 (0.7)^4 - 3.96665924 = 0$$
$$F_2 = b_0 + b_1 (1.6) + b_2 (1.6)^2 + b_3 (1.6)^4 - 2.06200548 = 0$$
$$F_3 = b_0 + b_1 (2.2) + b_2 (2.2)^2 + b_3 (2.2)^4 + 0.233496575 = 0,$$

which can be solved for the unknown coefficients b_j.

Adopting the SIM.ROOTS format for expressing such equations, remembering that the coefficients b_0, b_1, b_2 and b_3 must be expressed as X(0), X(1), X(2) and X(3), we write:

```
10 F(0)=X(0)+X(1)*0.1+X(2)*0.1↑2+X(3)*0.1↑4–3.56482944:RETURN
11 F(1)=X(0)+X(1)*0.7+X(2)*0.7↑2+X(3)*0.7↑4–3.96665924:RETURN
12 F(2)=X(0)+X(1)*1.6+X(2)*1.6↑2+X(3)*1.6↑4–2.06200548:RETURN
13 F(3)=X(0)+X(1)*2.2+X(2)*2.2↑2+X(3)*2.2↑4+0.233496575:
RETURN
```

RUNning the program and providing the initial parameters shown below, we obtain:

```
NR OF EQUATIONS? 4
CONSTANTS? (Y/N) N
DECIM ACCUR = .0001
NR OF ITERATIONS 100
OUTPUT RESULTS TO PRINTER? (Y/N) N
X(0) = 0
X(1) = 0
X(2) = 0
X(3) = 0
R(0) = 0
R(1) = 0
R(2) = 0
R(3) = 0
```

```
RE-ARRANGE EQUATIONS? (Y/N) Y
ENTER EQUATION SEQUENCE
ORIGINAL F()        NEW SEQUENCE
0                   0
1                   1
2                   3
3                   2

SIGN SEARCH? (Y/N) N
SIGN F(0) = -1
SIGN F(1) = -1
SIGN F(2) = 1
SIGN F(3) = -1

SIGN COMB.-1-1 1 -1
EQU. SEQU. 0  1  3  2

ITER   P    R    ROOT
1      1    0    1.98337964
1      1    0    1.43613278
1      0    1    -.945971623
1      0    1    -.0798231371
2      2    0    3.45140549
2      1    1    1.43245903
2      1    1    -1.1353624
2      0    2    -.0303966562
3      2    1    3.43977374
3      1    2    1.45437844
3      2    1    -1.52866656
3      0    3    -.0698124513
4      3    1    3.43573502
4      2    2    1.54105524
4      2    2    -1.07754729
4      0    4    -.0406060419
-      -    -    ____________
-      -    -    ____________
-      -    -    ____________
-      -    -    ____________
79     60   19   3.34779568
79     60   19   2.38848736
79     57   22   -2.18547104
79     58   21   .0742548824

SOLUTION                        RESIDUAL
X(0) =3.34779568            F(0)=-3.231E-05
X(1)=2.38848736             F(1)=2.538E-05
```

X(2)=-2.18547104 F(2)=-7.991E-04
X(3)=.0742548824 F(3)=-2.250E-03

SIGN COMB.-1-1 1-1
EQU. SEQU. 0 1 3 2

TO CONTINUE, TYPE ... CONT

Substituting the values for X(0), X(1), X(2) and X(3) into Equation (6.3), we obtain the required polynomial:

$$f(t)=3.34779568+2.38848736\,t-2.18547104\,t^2+0.074258824\,t^4.$$

We can use the ROOTS program to find the root of the above polynomial which is at t=2.14196955. This compares well with the quoted root of the generating function at t=2.14159265.

The greatest error between the two curves, over the given range, is when

$$\frac{d}{dt}\left[f(t)-4\sin(t+1)\right]=0,$$

which itself can be solved by the ROOTS program. There are three roots of this expression within the given range, the largest residual of which is the maximum error, namely 0.014. Check, as shown in Section 6.4, that the polynomial fits the data points.

For interest only, we can calculate and compare the area under the data-generating sine curve with that of the quartic polynomial with limits of integration covering the data range (in this case from t=0.1 to t=2.2). Integrating analytically, the area under the sine curve is found to be 5.80756359 and that under the polynomial, 5.80773633, which corresponds to an error of 0.003%.

Simpson's rule can also be applied to automatically integrate the curve (see utility program in Appendix E).

6.2 CUBIC SPLINES

Fitting a polynomial of high degree through a set of more than six data points is a tedious process. An alternative method is to fit a set of cubics that pass through the data points, using a new cubic in each interval, as shown in Fig. 6.2.

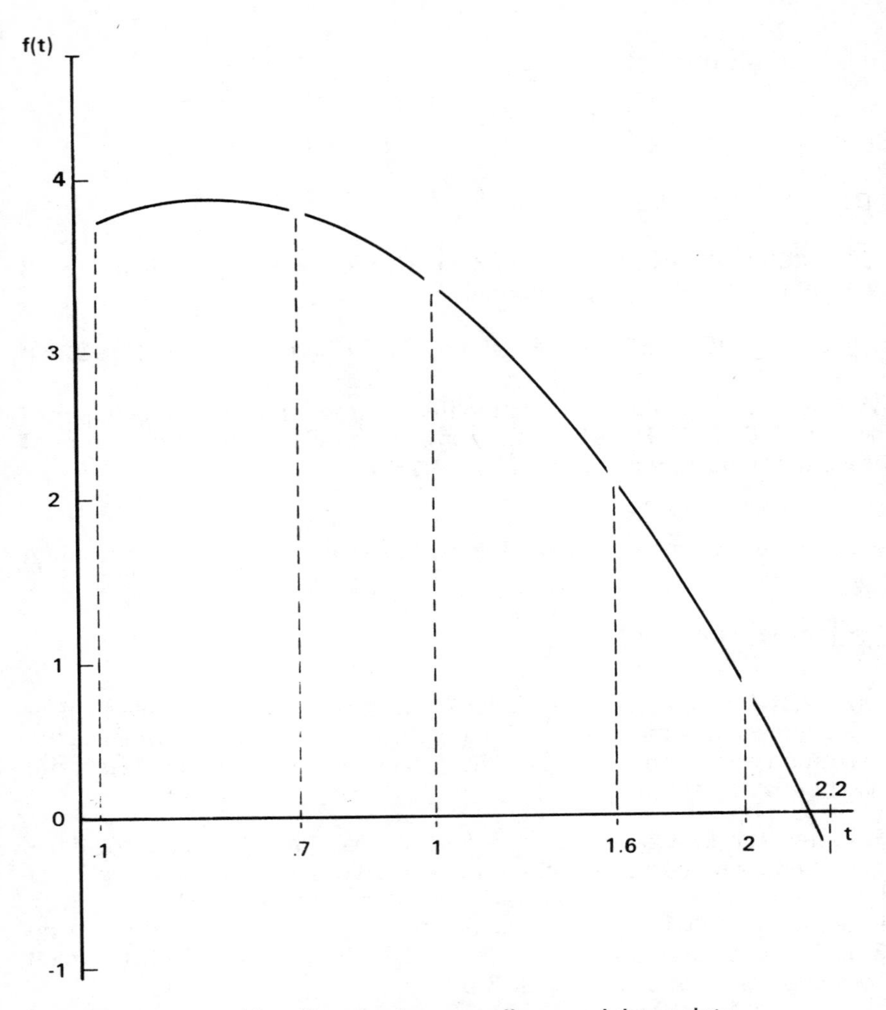

Figure 6.2 Fitting cubic splines to six unequally spaced data points.

The condition for a cubic spline-fit is that both the slope $f'(t)$ and the curvature $f''(t)$ (i.e. both the first and second derivatives), be the same for the pair of cubics that join at each interval, similar to a draughtsman's spline or flexy curve.

Thus, we can write the cubic of the ith interval which passes through the points $t_i, f(t_i)$ and t_{i+1}, $f(t_{i+1})$ in the form:

$$f(t_i) = a_i + b_i\,(t-t_i) + c_i\,(t-t_i)^2 + d(t-t_i)^3. \quad (6.4)$$

What follows is only necessary in order to derive the set of equations, the solution of which will produce the required cubic splines. Once the equations have been derived, the method of fitting the cubic splines to data is quite simple and straightforward.

Now, since the cubic in the ith interval is required to fit at the two end points t_i, $f(t_i)$ and t_{i+1}, $f(t_{i+1})$, we obtain (after substitution into Equation 6.4):

$$f(t_i) = a_i \quad (6.5)$$

$$f(t_{i+1}) = a_i + b_i\,(t_{i+1}-t_i) + c_i\,(t_{i+1}-t_i)^2 + d_i\,(t_{i+1}-t_i)^3,$$

or writing h_i for $(t_{i+1}-t_i)$,

$$f(t_{i+1}) = a_i + b_i\,h_i + c_i\,h_i^2 + d_i\,h_i^3. \quad (6.6)$$

We need the first and second derivatives of the spline in order to relate the slope and the curvature of the adjoining splines. Thus, differentiating Equation (6.4) twice, we obtain:

$$f'(t_i) = b_i + 2c_i\,(t-t_i) + 3d_i\,(t-t_i)^2 \quad (6.7)$$

$$f''(t_i) = 2c_i + 6d_i\,(t-t_i) \quad (6.8)$$

and substituting $t=t_i$ and $t=t_{i+1}$ into Equation (6.8), we get:

$$f''(t_i) = 2c_i \quad (6.9)$$

$$f''(t_{i+1}) = 2c_i + 6d_i\,h_i. \quad (6.10)$$

Substituting Equation (6.9) into Equation (6.10) and solving for d_i, we obtain:

$$d_i = \frac{f''(t_{i+1}) - f''(t_i)}{6h_i} \quad (6.11)$$

Finally, substituting a_i, c_i and d_i (given by Equations (6.5), (6.9) and (6.11)) into Equation (6.6), gives b_i, as follows:

$$b_i = \frac{f(t_{i+1}) - f(t_i)}{h_i} - \frac{[2f''(t_i) + f''(t_{i+1})]h_i}{6} \qquad (6.12)$$

Now, since the slope of the two cubics that join at the point t_i, $f(t_i)$ are the same, we can write:

$$f'(t_{i-1}) = f'(t_i) \qquad (6.13)$$

and from Equation (6.7) we obtain:

$$f'(t_{i-1}) = b_{i-1} + 2c_{i-1}h_{i-1} + 3d_{i-1}h_{i-1}^2, \text{ and}$$

$$f'(t_i) = b_i.$$

Equating these and substituting for the appropriate expressions for b, c and d, we get:

$$f'(t_{i-1}) = \frac{f(t_i) - f(t_{i-1})}{h_{i-1}} - \frac{[2f''(t_{i-1}) + f''(t_i)]\,h_{i-1}}{6}$$

$$+ f''(t_{i-1})\,h_{i-1} + \frac{[f''(t_i) - f''(t_{i-1})]\,h_{i-1}}{2}$$

$$= \frac{f(t_{i+1}) - f(t_i)}{h_i} - \frac{[2f''(t_i) + f''(t_{i+1})]\,h_i}{6}$$

This can be simplified to give the recurrence relationship:

$$h_{i-1}f''(t_{i-1}) + [2h_{i-1} + 2h_i]\,f''(t_i) + h_i f''(t_{i+1})$$

$$-6\left[\frac{f(t_{i+1}) - f(t_i)}{h_i} - \frac{f(t_i) - f(t_{i-1})}{h_{i-1}}\right] = 0 \qquad (6.14)$$

which applies at each interval point from i=1 to i=n-1 for a total of n+1 points (i.e. n-1 equations relating n+1 points).

We can obtain two extra equations for the first and last points by specifying the end conditions of the whole curve which are largely arbitrary. If $f''(t_0)$ is a linear extrapolation from $f''(t_1)$ and $f''(t_2)$ (with similar conditions applied to $f''(t_n)$ relative to $f''(t_{n-1})$ and $f''(t_{n-2})$), we find that for a set of data, fitted through by a single cubic, the spline curve is this same cubic. Alternative end conditions don't have this property.

Thus, assuming linearity at the end points, we obtain:

$$\frac{f''(t_1) - f''(t_0)}{h_0} = \frac{f''(t_2) - f''(t_1)}{h_1}$$

and

$$\frac{f''(t_n) - f''(t_{n-1})}{h_{n-1}} = \frac{f''(t_{n-1}) - f''(t_{n-2})}{h_{n-2}}$$

or

$$h_1 f''(t_0) - (h_0 + h_1) f''(t_1) + h_0 f''(t_2) = 0$$

and

$$h_{n-1} f''(t_{n-2}) - (h_{n-2} + h_{n-1}) f''(t_{n-1}) + h_{n-2} f''(t_n) = 0. \quad (6.15)$$

Hence, we can write a set of equations (utilizing Equation (6.15) for the end intervals and Equation (6.14) for the inner intervals) to fit cubics to any number of data points with the quantities $f''(t_i)$ being the unknowns. Writing these as x_i, since unknowns must be expressed in this form for the eventual use of the SIM.ROOTS program, we can write the required equations for a typical six point problem, in the following way:

$$F_0 = h_1 x_0 - (h_0 + h_1) x_1 + h_0 x_2 = 0$$

$$F_1 = h_0 x_0 + 2(h_0 + h_1) x_1 + h_1 x_2 - 6 \left[\frac{f(t_2) - f(t_1)}{h_1} - \frac{f(t_1) - f(t_0)}{h_0} \right] = 0$$

$$F_2 = h_1 x_1 + 2(h_1 + h_2) x_2 + h_2 x_3 - 6 \left[\frac{f(t_3)-f(t_2)}{h_2} - \frac{f(t_2)-f(t_1)}{h_1} \right] = 0$$

$$F_3 = h_2 x_2 + 2(h_2 + h_3) x_3 + h_3 x_4 - 6 \left[\frac{f(t_4)-f(t_3)}{h_3} - \frac{f(t_3)-f(t_2)}{h_2} \right] = 0$$

$$F_4 = h_3 x_3 + 2(h_3 + h_4) x_4 + h_4 x_5 - 6 \left[\frac{f(t_5)-f(t_4)}{h_4} - \frac{f(t_4)-f(t_3)}{h_3} \right] = 0$$

$$F_5 = h_4 x_3 - (h_3 + h_4) x_4 + h_3 x_5 = 0 \qquad (6.16)$$

These equations would be radically simplified were we to take equally spaced data in which case all h_i would be the same. Nevertheless, with the help of a tabulation procedure (to be introduced shortly), even unequally spaced data can be handled easily.

Note that the resultant equations prove to be well-behaved and their solution is straight forward. As an example, we will consider the same four coordinate problem we discussed earlier, but with the addition of two extra data points.

6.3 PROCEDURE FOR FITTING CUBIC SPLINES

Let us assume that we are required to fit cubic splines as shown in Fig. 6.2, through the following six data points:

ti	f(ti)
0.1	3.56482944
0.7	3.96665924
1.0	3.6371897
1.6	2.06200548
2.0	0.564480028
2.2	−0.233496575

In order to simplify and minimize the amount of work in entering Equations (6.16) into the computer, we suggest that Table 6.1 be formed and filled in. The Table deals, significantly, with the calculation of the final constant term in each of the internal interval equations.

TABLE 6.1 Data for Spline Fitting

i	t_i	h_i	$f(t_i)$	$f(t_{i+1})$-$f(t_i)$	$-6*\triangle f(t_i)/h_i$
0	0.1	0.6	3.56482944	+0.4018298	
1	0.7	0.3	3.96665924	−0.32946953	10.6076886
2	1.0	0.6	3.63718971	−1.57518423	9.16245168
3	1.6	0.4	2.06200548	−1.49752545	6.71103947
4	2.0	0.2	0.564480028	−0.797976603	1.47641633
5	2.2		−0.233496575		

We are now in a position to type Equations (6.16) into the SIM.ROOTS program, as follows:

```
10 F(0)=0.3*X(0)–(0.6+0.3)*X(1) + 0.6*X(2):RETURN
11 F(1)=0.6*X(0) + 2*(0.6+0.3)*X(1) + 0.3*X(2) + 10.6076886:
   RETURN
12 F(2)=0.3*X(1) + 2*(0.3+0.6)*X(2) + 0.6*X(3) + 9.16245168:
   RETURN
13 F(3)=0.6*X(2) + 2*(0.6+0.4)*X(3) + 0.4*X(4) + 6.71103947:
   RETURN
14 F(4)=0.4*X(3) + 2*(0.4+0.2)*X(4) + 0.2*X(5) + 1.47641633:
   RETURN
15 F(5)=0.2*X(3)–(0.4+0.2)*X(4) + 0.4*X(5):RETURN
```

RUN the program and provide the following initial parameters:

```
NR OF EQUATIONS? 6
CONSTANTS? (Y/N) N
DECIM ACCUR = .0000001
```

```
NR OF ITERATIONS 100
OUTPUT RESULTS TO PRINTER? (Y/N) N
X(0) = 0
X(1) = 0
X(2) = 0
X(3) = 0
X(4) = 0
X(5) = 0
R(0) = 0
R(1) = 0
R(2) = 0
R(3) = 0
R(4) = 0
R(5) = 0
RE-ARRANGE EQUATIONS? (Y/N) N
SIGN SEARCH? (Y/N) Y

SIGN COMB.-1 -1 -1 -1 -1 -1
EQU. SEQU. 0  1  2  3  4  5
ITER  P   R   ROOT
1     1   0    0
1     0   1   -1.2131477
1     0   1   -1.13933454
1     0   1   -.990649671
1     0   1   -.371821928
1     0   1   -9.9056329E-03
2     2   0   -.501030086
2     1   1   -2.58765419
2     1   1   -2.36347184
2     1   1   -1.9250233
2     1   1   -.499548027
2     0   2    7.77803068E-03
3     3   0   -1.61462898
3     2   1   -3.94753467
3     2   1   -3.41710004
3     2   1   -2.28924172
3     1   2   -.490254571
3     1   2    .0477535565
4     4   0   -3.40290348
4     3   1   -4.28292526
4     3   1   -3.69194789
4     2   2   -2.20255077
4     1   3   -4.92334631
4     2   2    .0873392848
```

```
-      -      -     ____________
-      -      -     ____________
-      -      -     ____________
-      -      -     ____________
-      -      -     ____________
-      -      -     ____________
40     31     9     -4.16987933
40     29     11    -3.99055239
40     30     10    -3.73588889
40     28     12    -2.12280996
40     30     10    -.559715562
40     31     9     .221831636

SOLUTION                          RESIDUAL

X(0)=-4.16987933                  F(0)=1.8652E-08
X(1)=-3.88055239                  F(1)=4.0047E-08
X(2)=-3.73588889                  F(2)=-6.519E-09
X(3)=-2.12280996                  F(3)=-1.863E-09
X(4)=-.559715562                  F(4)=0
X(5)=.221831636                   F(5)=2.1464E-10

SIGN COMB.-1 -1 -1 -1 -1 -1
EQU. SEQU. 0  1  2  3  4  5

TO CONTINUE, TYPE ... CONT
```

Now, the coefficients a_i, b_i, c_i, and d_i can be found from Equations (6.5), (6.12), (6.9) and (6.11), respectively, for each spline interval. For the sake of easy reference, we reproduce these equations below.

$$a_i = f(t_i)$$

$$b_i = \frac{f(t_{i+1}) - f(t_i)}{h_i} - \frac{\left[2f''(t_i) + f''(t_{i+1})\right] h_i}{6}$$

$$c_i = \frac{f''(t_i)}{2}$$

$$d_i = \frac{f''(t_{i+1}) - f''(t_i)}{6h_i}$$

Thus, using the above equations and Table 6.1, we can produce Table 6.2.

Alternatively, as there is rather a lot of arithmetic associated with the calculation of b_i, c_i, and d_i, we could write the above equations into the SIM.ROOTS program and let the computer do the work for us. To do this, type the following substitution equations with their associated line numbers into the computer.

```
16 F(6)=X(6)—0.4018298/0.6 + (2*X(0)+X(1))*0.6/6:RETURN
17 F(7)=X(7)—X(0)/2:RETURN
18 F(8)=X(8)—(X(1)-X(0))/(6*0.6):RETURN
19 F(9)=X(9) + 1.57518423/0.6 + (2*X(2)+X(3))*0.6/6:RETURN
20 F(10)=X(10)—X(2)/2:RETURN
21 F(11)=X(11)—(X(3)-X(2))/(6*0.6):RETURN
22 F(12)=X(12) + 1.49752545/0.4 + (2*X(3)+X(4))*0.4/6:RETURN
23 F(13)=X(13)—X(3)/2:RETURN
24 F(14)=X(14)—(X(4) —X(3))/(6*0.4):RETURN
```

On RUNning the program, solutions will be reached in the same number of steps as before, but with the additional results X(6) to X(14), as shown below:

SOLUTION	VALUE OF	COEFFICIENTS OF
X(6)=1.89174744	b_1	
X(7)=-2.08493966	c_1	1 st spline
X(8)=.0803686013	d_1	
X(9)=-1.66584828	b_2	
X(10)=-1.86794444	c_2	2 nd spline
X(11)=.448077479	d_2	
X(12)=-3.42345793	b_3	
X(13)=-1.06140498	c_3	3 rd spline
X(14)=.651289332	d_3	

With the above values in hand, we can now construct Table 6.2, as follows:

TABLE 6.2 Coefficients of Cubics for the Three Spline Intervals

t_i =	spline interval 0.1-0.7-10	spline interval 1.0 - 1.6	spline interval 1.6-2.0-2.2
Coeff			
a_i	3.56482944	3.63718971	2.06200548
b_i	1.89174744	-1.66584828	-3.42345793
c_i	-2.08493966	-1.86794444	-1.06140498
d_i	0.0803686013	0.448077479	0.651289332

Substituting the above coefficients back into Equation (6.4), we obtain the three cubic splines which cover the whole data range from t=0.1 to t=2.2, as shown below:

$$f(t_{0.1-1.0})=3.56482944 + 1.89174744\,(t-0.1)-2.08493966\,(t-0.1)^2 + 0.0803686013\,(t-0.1)^3$$

$$f(t_{1.0-1.6})=3.63718971 - 1.66584828\,(t-1.0) - 1.86794444\,(t-1.0)^2 + 0.448077479\,(t-1.0)^3$$

$$f(t_{1.6-2.2})=2.06200548 - 3.42345793\,(t-1.6) - 1.06140498\,(t-1.6)^2 + 0.651289332\,(t-1.6)^2$$

6.4 CHECKING SPLINE AND DATA COINCIDENCE

You can use the ROOTS program to check whether each of the above splines passes through all the data points of its respective interval. To do this, type each cubic spline into the computer in line 10 (remembering to substitute X for t), and use the 'Fine Search' facility with an increment of 0.1. As an example, we will perform this with the third spline, as shown below:

```
ACCUR. = .0000001
NR OF ITERATIONS? 50
X = 0
R = 0
FINE SEARCH? (Y/N) Y
SEARCH INCREMENT? .1
SIGN = 1


F(X)                                X

2.61092968                         .1
2.98735489                         .2
3.28784371                         .3
3.51630386                         .4
3.67664307                         .5
3.7727691                          .6
3.80858966                         .7
3.7880125                          .8
3.71494535                         .9
3.59329595                         1
```

3.42697203	1.1
3.21988134	1.2
2.9759316	1.3
2.69903055	1.4
2.39308593	1.5
2.06200548	1.6
1.70969692	1.7
1.340068	1.8
.957026459	1.9
.564480024	2
.166336435	2.1
-.233496577	2.2

ITER	P	R	ROOT
1	0	0	2.01631764
2	0	1	2.20721252
3	0	2	2.15574552
4	1	2	2.14160446
5	2	2	2.14159477
6	2	3	2.14159603
7	3	3	2.14159662
8	4	3	2.14159677
9	4	4	2.14159677

X=2.14159677 F=-1.15460352E-09

TO CHANGE PARAMETERS, TYPE ... RUN

The function values at t=1.6, 2.0 and 2.2 should exactly equal the original data values. Further, note the value of the root and compare this with the true position of the root at t=2.14159265 as given by the generating function.

As a final check, we could use the SIMPSON's rule program (Appendix E) to calculate the area under each spline and then compare the cumulative value with the known area under the curve, just as we did in the case of the polynomial fit.

As always, decreasing the intervals h_i, at the expense of increasing the number of equations, will result in greater precision in estimating the ideal curve, provided of course the extra data points are experimentally available. Do not attempt to extrapolate data outside a spline interval, as this has an adverse effect on accuracy.

6.5 FITTING NON-POLYNOMIAL FUNCTIONS TO DATA

In addition to polynomials, other functions or series, such as trigonometric, exponential or asymptotic, can be fitted to data points. Thus, the asymptotic series

$$f = a_0 + a_1/t + a_2/t^2 \qquad (6.17)$$

can be fitted to the data points (1/3,0), (1/2,–2) and (1,0) by substituting for t into the above equation.
A sketch of the function is shown in Fig. 6.3.

Replacing a_i by X_i, we obtain the familiar SIM.ROOTS format:

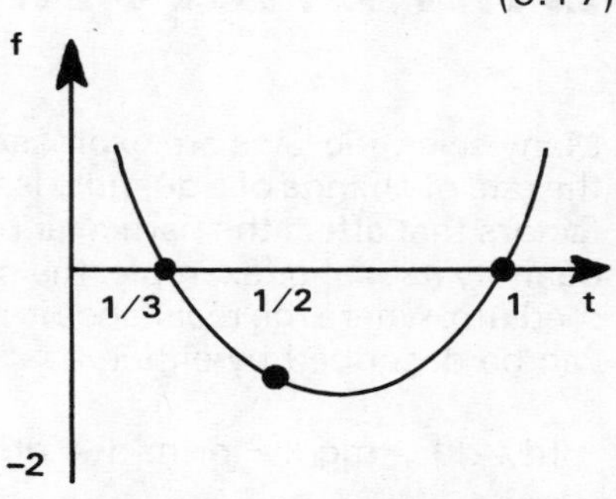

Figure 6.3 Asymptotic function.

F(0)=X(0) + X(1)/(1/3) + X(2)/(1/3)↑2 = 0
F(1)=X(0) + X(1)/(1/2) + X(2)/(1/2)↑2+ 2 = 0
F(2)=X(0) + X(1) + X(2) = 0.

Typing the above three equations into the program and providing the initial parameters:

DECIM ACCUR = .001
All Xs = .5
All Rs = 0
EQU. SEQU. 2 0 1
SIGN COMB. 1 –1 –1

the following solutions are reached in 94 steps:

$X(0) = a_0 = 5.99$
$X(1) = a_1 = -7.99$
$X(2) = a_2 = 1.99.$

CHAPTER 7

INITIAL-VALUE DIFFERENTIAL EQUATIONS

Many scientific laws are expressed in the form of differential equations if the rate of change of a quantity is more easily determined from the various factors that affect the particular phenomenon than the actual value of the quantity itself. For example, the motion of a free falling body in a resistive medium, where the resistance is proportional to the square of the velocity, can be described by either

$$m(dv/dt) = mg - kv^2 \text{ or } m(d^2s/dt^2) = mg - k(ds/dt)^2,$$

which, after rearranging, can be written as

$$v' + (k/m)v^2 - g = 0 \text{ and } s'' + (k/m)s'^2 - g = 0,$$

where $v' = (dv/dt)$ = the rate of change of velocity with time and $s' = (ds/dt)$ = rate of change of distance with time = velocity. Thus, in the above expression, $v' \equiv s''$.

Differential equations are classified by the highest order derivative present. Thus, the first equation is classified as a first order differential equation, connecting velocity v and time, while the second is classified as a second order differential equation, connecting distance s and time.

The analytic solution to a differential equation (Lamb, 1956) is the function which satisfies both the differential equation and certain initial conditions on the function. So in the second of the above examples, our solution would be an equation in s as a function of t. The numerical solution to the same differential equation, on the other hand, involves a tabulation of the values of the function at various values of the independent variable (Gerald, 1970). For the first order differential equation of the above example, we would be tabulating or plotting values of v at various values of t.

Traditional methods of solving differential equations often demand a high level of skill and expertise which inhibits their use. In this Chapter we will formulate a method which uses the same algorithm and essentially the same program as the one used to solve simultaneous equations. The method is capable of solving linear or very complicated, implicit, transcendental and simultaneous differential equations.

7.1 NUMERICAL STEPWISE INITIAL-VALUE METHODS

Most numerical stepwise-approach methods of solving say, an initial-value third order differential equation, such as

$$F(x''', x'', x', x, t) = 0 \qquad (7.1)$$

consist in starting at one known point $t=t_0$, $x=x_0$ at which all the initial lower differential orders are known, then progressing along the t-axis by a small increment h, and recording or plotting the corresponding points given by the differential equation algorithm. These points should then represent a very close approximation to the primitive equation curve, provided the function and all its appropriate derivatives exist and are continuous over the relevant range of t.

Many iterative algorithms exist for reproducing the differential equation solution points. Some, like Euler's (Gerald, 1970 and Shampine, 1973), are easy to apply, but are of very low precision; others, such as the Runge-Kutta (Gerald, 1970 and Stanton, 1961) or various Predictor-Corrector (Noble, 1972) schemes, are very precise, yet of limited or non-existent application, when trying to solve high order simultaneous differential equations, as they are not always convergent.

Some algorithms, such as the Fox-Goodwin (Watson 1974), are quite laborious, especially in starting, while others, including the above, are of specific application and hard to computerize. Predictor-Corrector techniques usually need 2–5 starting points which are derived by yet another technique, resulting in a rather involved method.

7.2 ADOPTION OF TAYLOR'S SERIES METHOD

The Taylor series method of solving differential equations is based on calculating the coefficients of the Taylor series

$$x(t+h) = x(t) + hx'(t) + \frac{h^2}{2!}x''(t) + \frac{h^3}{3!}x'''(t) + \frac{h^4}{4!}x^{IV}(t) + \ldots \qquad (7.2)$$

If we know x_r and t_r then we can calculate x_{r+1} at $t_{r+1} = t_r + h$, since the derivatives x'_r, x''_r, ..., can be obtained from the problem equation such as Equation (7.1). The method has great stability and its accuracy can be impoved by simply adding more terms or decreasing the increment h. Furthermore, the method is flexible in allowing the user to change parameters and easily interlocks with our algorithm.

The disadvantage of the method lies in the occasional need to differentiate possibly difficult functions, though a substitution arrangement can frequently reduce the involved labour as will be explained later.

The incorporation of the Taylor's series method into the SIM.ROOTS program, results in a very powerful method for easily solving even highly non-linear differential equations. Convergency checking is minimized since each derived (differentiated) function can, in the worst case, be first solved for its unique highest differential prior to final substitution into the basic Taylor's series formula. Once convergency is reached, such partial solutions and substitutions are all performed iteratively by the program. Thus, the first level function (the actual problem equation) is the Euler approximation, while subsequent derived differential equations add to the precision of the answer.

In this text there is no need to distinguish between the traditional categories of differential equations, such as implicit, explicit, linear, non-linear, well- or ill-conditioned, transcendental or homogeneous, because the algorithm is impartial in its treatment of differential equation types. However, the function must be analytic (regular) in the region of interest.

7.3 SOLVING SINGLE DIFFERENTIAL EQUATIONS USING SIM.ROOTS

Consider the second order differential equation problem function F(2) – [a third order would be written as F(3)] – given by

$F(2) = X\,X'' + 1 + (X')^2 = 0,$

where $X'' = \frac{d^2X}{dT^2}$ and $X' = \frac{dX}{dT}$

with variable T having any physical meaning assigned to it. Then, assuming we desire high precision, we differentiate F(2) until we reach F(5) which will contain maximum differential orders of function X, up to

$$X^{V} = \frac{d^5X}{dT^5}$$

Thus, we obtain the following equations:

$$F(2) = X\,X'' + 1 + (X')^2 = 0$$
$$F(3) = F'(2) = X\,X''' + 3\,X'\,X'' = 0$$
$$F(4) = F'(3) = 4\,X'\,X''' + X\,X^{IV} + 3\,(X'')^2 = 0$$
$$F(5) = F'(4) = 5\,X'\,X^{IV} + 7\,X''\,X''' + X\,X^{V} = 0$$

and after substituting the relevant constants C(0) and C(1) for X and X′, we can write:

$$\begin{aligned} F(2) &= C(0)\,X'' + 1 + (C(1))^2 = 0 \\ F(3) &= C(0)\,X''' + 3\,C(1)\,X'' = 0 \\ F(4) &= 4\,C(1)\,X''' + C(0)\,X^{IV} + 3\,(X'')^2 = 0 \\ F(5) &= 5\,C(1)\,X^{IV} + 7\,X''\,X''' + C(0)\,X^{V} = 0. \end{aligned} \qquad \text{... (7.3)}$$

The constants C(0) and C(1) in the above equations are dictated as initial starting conditions of the problem as follows:

$T_{old} = 0$

$X = C(0)_{old} = 1$

$X' = C(1)_{old} = 0.$

It is easy to see that the problem F(2) is in reality a circle of radius C(0)=1 (Fig. 7.1) in which $X^2 = 1 - T^2$.

With the starting conditions T, C(0) and C(1) substituted into Equations 7.3, we have four simultaneous equations (very elementary in this case) which can be solved for X'', X''', X^{IV} and X^{V} (or X(2), X(3), X(4) and X(5) in the SIM.ROOTS program terminology) using the computer.

Note: Always check that the number of unknowns X(2) to X(5) equals the number of equations and that the number of constants equals the differential order of the problem equation.

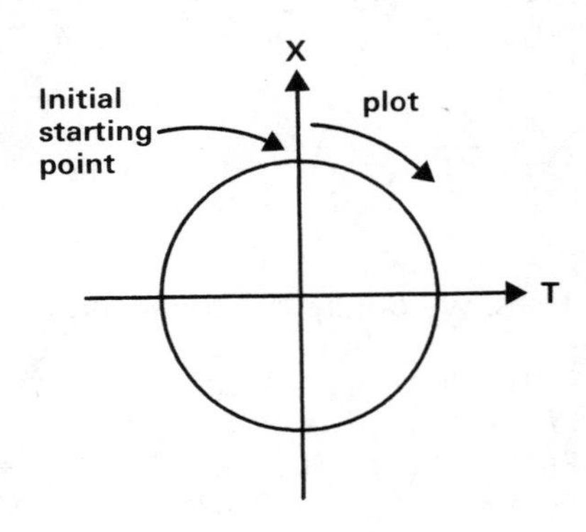

Figure 7.1 Tracking a circle.

Very rarely, more difficult problems could result in an implicit or transcendental equation such as

$$F(2) = X^{T}\,[\cos(1+X'')]^{1/2} + (X'')^{2}\,X' - T^{X} = 0,$$

which is as easily handled by the SIM.ROOTS program in evaluating X'' etc. However, the problem equation with its derived equations could have several roots, so the operator should be careful on these rare occasions.

Returning to the previous example, we can mentally solve the simultaneous Equations 7.3, to obtain:

$X(2) = X'' = -1$
$X(3) = X''' = 0$
$X(4) = X^{IV} = -3$
$X(5) = X^{V} = 0.$

In order to obtain the final answers we need to substitute the above values together with the values of the two initial constants, $C(0)_{old}$, $C(1)_{old}$, into the two Taylor's equations given below (three constants into three Taylor's series are required for a third order differential equation), to obtain:

$$X = X(0) = C(0)_{old} + H\,C(1)_{old} + \frac{H^2}{2!}X'' + \frac{H^3}{3!}X''' + \frac{H^4}{4!}X^{IV} + \frac{H^5}{5!}X^{V}$$

$= C(0)_{new}$ next time around, and

$$X = X(1) = C(1)_{old} + H\,X'' + \frac{H^2}{2!}X''' + \frac{H^3}{3!}X^{IV} + \frac{H^4}{4!}X^{V}$$

$= C(1)_{new}$ next time around.

Letting H = .1 (i.e. small increment), we now jump from

$T = 0,\ C(0)_{old} = 1,\ C(1)_{old} = 0$

to

$T = T + H = 0 + .1 = .1,\ C(0)_{new} = X,\ C(1)_{new} = X$

where $C(0)_{new} = .9949875$ and $C(1)_{new} = -.1005$.

All equations derived from the problem are said to be in one sub-set. Thus, there are four equations (not counting Taylor's) in this sub-set. At much less accuracy there could be only one equation in the sub-set, the problem.

With new values for C(0) = X(0) and C(1) = X(1), substituted back into Equations 7.3, the whole procedure is repeated methodically, in order to find the next point. If you are using a calculator you will find it easier if you devise a results table.

Alternatively, you could use the SIM.ROOTS program to simultaneously solve Equations 7.3 and type the two Taylor's series equations as substitution equations for the required final substitution. The whole procedure would then have to be repeated for each plotted point, remembering that the results of the last program execution must be used as the input to the new program execution.

Changing all Gs to Cs in line 145 of the SIM.ROOTS program and adding two more statements at the end of it, as shown below, will allow the insertion of all Taylor's constants to be made easily. Note that the problem equation and its derived equations could contain functions of T.

```
145 INPUT "CONSTANTS? (Y/N) ";T$:IF LEFT$(T$,1)="Y" THEN
    INPUT "HOW MANY? ";C:DIM C(C):FOR I=0 TO C-1:
    PRINT "C(";I;")=";:INPUT "";C(I):NEXT I:
    INPUT "H=";H:INPUT "T=";T
```

Writing the equations to be solved in the SIM.ROOTS format, starting with the two Taylor's series equations, we obtain:

```
10 F(0)=X(0) – (C(0) + H*C(1) + H↑2/2*X(2) + H↑3/6*X(3) +
   H↑4/24*X(4) +1 H↑5/120*X(5)):RETURN
11 F(1)=X(1) – (C(1) + H*X(2) + H↑2/2*X(3) + H↑3/6*X(4) +
   H↑4/24*X(5)):RETURN
12 F(2)=C(0)*X(2) + 1 + C(1)↑2:RETURN
13 F(3)=C(0)*X(3) + 3*C(1)*X(2):RETURN
14 F(4)=4*C(1)*X(3) + C(0)*X(4) + 3*X(2)↑2:RETURN
15 F(5)=5*C(1)*X(4) + 7*X(2)*X(3) + C(0)*X(5):RETURN
```

RUN the program and provide the following initial parameters:

```
NR OF EQUATIONS? 6
CONSTANTS? (Y/N) Y
HOW MANY? 2
C(0)=1
C(1)=0
H=.1
T=0
DECIM ACCUR = .000000001
NR OF ITERATIONS 100
OUTPUT RESULTS TO PRINTER? (Y/N) N
```

and setting all Xs = 0, Rs = 0 and SIGNs =–1, we obtain (after 17 iterations) the following SIM.ROOTS results:

SOLUTION	RESIDUAL
X(0)=.9949875	F(0)=0
X(1)=–.1005	F(1)=–2.9E–11
X(2)=–1	F(2)=0
X(3)=0	F(3)=0
X(4)=–3	F(4)=–9.3E–10
X(5)=0	F(5)=0

SIGN COMB. –1 –1 –1 –1 –1 –1
EQU. SEQU. 0 1 2 3 4 5

Remember that the first two answers have the following meaning:

X(0) = X
X(1) = X′

and that for the next step, we require to change the initial parameters to:

C(0) = X(0)
C(1) = X(1)
H=.1
T=.1

in order to obtain the second plotted point.

The circle differential equation (Equations 7.3) is ideal for beginners to the above procedure because it allows the solution of a differential equation around a quarter circle, a curve whose slope is continually increasing until, near X = 1, it approaches infinity and hence is liable to grater inaccuracy, unless the increment H is reduced drastically as the slope increases. It is instructive to plot in the above fashion up to T = .8 and then, by setting H to similar but negative values, return to the starting point. Using the SIM.ROOTS program to carry out the above suggestion with constant H = .1 increments, the result

$$X = X(0) = C(0)_{new} = .600240074$$

was obtained at T = .8, whereas it should have been .6 exact.

Note that convergency can be tested, mistakes eliminated and stability assured before actually performing the first plot by setting H to zero and checking that the first F(2) equation solves satisfactorily alone; then solving equations F(2) and F(3) together, followed by F(2), F(3) and F(4), prior to solving all of them simultaneously.

7.3 SOLVING SIMULTANEOUS DIFFERENTIAL EQUATIONS

The SIM.ROOTS program can be used to solve simultaneous differential equations provided the second problem equation with its own Taylor's series and derived equations are typed into lines which follow consecutively those of the first sub-set. In order to illustrate the method, we will solve the following problem.

Problem:

The two simultaneous differential equations given below, describe the bridge cable catenary shown in Fig. 7.2.

$$8\,(X'')^2 - 2 - Y\,X' = 0$$

$$(Y')^2 - 1 - (X')^2 = 0.$$

The corresponding primitive equations are given (for comparison purpose only) by:

$X = 2\cosh(T/2)$, and
$Y = 2\sinh(T/2)$.

Starting from point A,
where the initial parameters are:

$T=0$, $X=2$, $X'=0$ and $Y=0$,

plot the solution in incremental steps of 0.1, upto $T=0.4$.

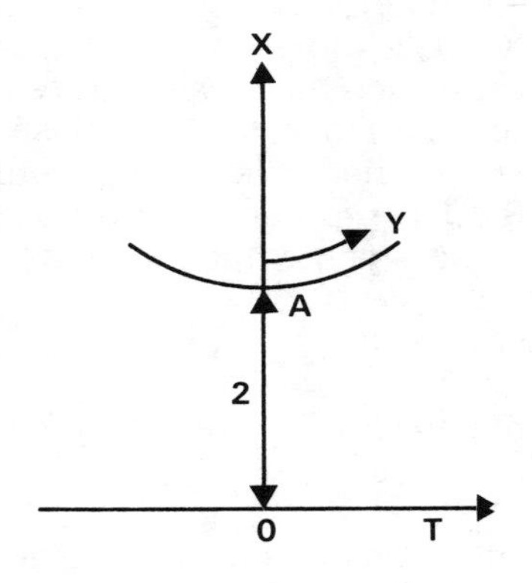

Figure 7.2 Tracking a catenary.

Solution:

The respective derived equations are:

$$16\,X''\,X''' - Y\,X'' - Y'\,X' = 0$$

$$2\,Y'\,Y'' - 2\,X'\,X'' = 0.$$

Formatting these to the requirements of the SIM.ROOTS program and preceding each with their appropriate Taylor's series equations, we obtain:

```
10 F(0)=X(0) – (C(0) + H*C(1) + H↑2/2*X(2) +
   H↑3/6*X(3)):RETURN
11 F(1)=X(1) – (C(1) + H*X(2) + H↑2/2*X(3)):RETURN
12 F(2)=8*X(2)↑2 – 2 – C(2)*C(1):RETURN
13 F(3)=16*X(2)*X(3) – C(2)*X(2) – X(5)*C(1):RETURN
14 F(4)=X(4) – (C(2) + H*X(5) + H↑2/2*X(6)):RETURN
15 F(5)=X(5)↑2 – 1 – C(1)↑2:RETURN
16 F(6)=2*X(6)*X(5) – 2*C(1)*X(2):RETURN
```

The first four lines form sub-set 1, while the last three form sub-set 2. Lines 10 and 11 are the Taylor's series equations of sub-set 1 (two equations since the problem equation is of the second order), while line 14 is the Taylor's series equation of sub-set 2. Note that the constants (Cs) in the Taylor's series equations must be suffixed consecutively in all sub-sets.

The initial parameters for the above simultaneous equations are 7 equations (the total number typed in) and 3 constants. Furthermore, since we start tracking from point A (see Fig. 7.2), then at T=0 we must take C(0)=2, C(1)=0, C(2)=0, all Xs=0, Rs=0 and SIGNs=–1. The answers at point T=.1 (implying an initial H=.1) are:

X(0)=C(0)=2.0025
X(1)=C(1)=0.05
X(4)=C(2)=0.1

After the solutions of the first plotted point are obtained, we must set the new values of C(0)=X(0), C(1)=X(1), C(2)=X(4), H=.1 (the same as before) and T= $(T_{old} + H)$=.1, before solving for the second plotted point.

The final results, at T=.4 are:

X(0)=C(0)=2.04011881
X(1)=C(1)=0.201251329
X(4)=C(2)=0.402502657.

We can see how closely these values conform with those given by the primitive equations presented earlier, by letting T=.4 and solving for X and Y, to obtain:

$X = 2 \cosh(0.4/2) = 2.040133$, and
$Y = 2 \sinh(0.4/2) = 0.402672$.

Note that accumulated round-off errors have affected the accuracy to some degree. Methods for improving the accuracy will be discussed in the next section.

The above scheme of entering equations requires the typing of Taylor's series equations each time the program is used to solve different sets of differential equations. It also requires the correct re-assignment of the constants before proceeding with the solution for the next plotted point. As these requirements can be programmed into the computer, we will introduce in the next chapter additions to the SIM.ROOTS program which will eliminate the above burden.

7.5 TESTING AND IMPROVING ACCURACY

As this is a practical book on solving equations, we are not including any theoretical error analysis which is adequately covered in other texts (Conte, 1972 and Gerald, 1970), particularly with regard to checking how Taylor's series progresses by summing odd and even terms. This procedure, amongst many, could be included in the new scheme in order to adjust H automatically should precision be threatened along the plot.

Another check involves looking at each of the two highest terms in Taylor's series for X, then multiplying the last two iteration results X_n and X_{n-1} by the appropriate $H^n/n!$ factor. Usually, these higher terms, when decreasing, indicate good convergency and precision.

Over-steepness of the slope $X' = X(1) = C(1)$, can illustrate a tendency towards instability, while doubling or halving H and checking how the value for $X = X(0) = C(0)$ varies for any given step is usually a very good method for gauging precision.

In general, there are several ways for improving the accuracy of plotting terms X relative to the unsolved primitive equation from which the differential equation is derived. These are:

(a) decreasing H, but only up to a point because more steps will cause the accumulation of round-off errors,

(b) adding more terms to the Taylor's series and therefore more derived equations to the sub-set at the expense of more effort,

(c) using a very high accuracy in order to reduce round-off errors,

(d) simplifying as many of the sub-set equations as possible,

(e) changing to polar co-ordinates (e.g. in the case of orbits) in order to avoid steep slopes, and

(f) inverting first order differential equations near a region of drastically increasing slope X'= X(1) = C(1) in order to prevent impending divergency. To illustrate this last point, assume that the equation

$$X' = T^2 + X$$

is to be solved subject to the initial condition that the solution passes through the point (0,10). Since X' is large, we can express the problem as

$$\frac{dT}{dX} = \frac{1}{T^2+X}$$

and obtain T as a Taylor series in X to assure convergency.

CHAPTER 8

DIFF.ROOTS PROGRAM FOR INITIAL-VALUE EQUATIONS

The SIM.ROOTS program can be converted to what we shall refer to as the DIFF.ROOTS program which can solve single or simultaneous initial-value differential equations (see Section 7.3) by the addition of the program lines listed in Appendix F. The program contains all the necessary Taylor's series equations, so that the user need only type in the problem and the derived equations.

The DIFF.ROOTS program automatically re-assigns the constants and re-circulates through the program solving for the next plotted point with several options being made available to the operator, such as ability to change the size of the step increment H and printing the results for future reference. The program must be re-LOADed before typing in new equations, as over-writing of the lines containing the Taylor's series would have taken place from previous use of the program.

For no special reason other than convenience, the computer program has been limited to simultaneously solving up to 4 differential equation functions X, Y, Z, W of T, each having a differential order anywhere from 0 to 4, i.e.,

X, X', X'', X''' or X^{IV}.

Each differential equation is associated with its own Taylor's series containing H power terms up to a maximum of H^5, depending on the operator's desire for precision and solution stability. The correct number of terms in each Taylor's series are taken automatically by the program (it depends on the number of derived equations) for maximum accuracy.

We shall now use the DIFF.ROOTS program to solve the circle differential equation (see Section 7.3). On typing 'RUN' the following information will appear on the screen:

```
FOR ROOTS OF SINGLE EQU. USE LINE 10

FOR SIMULTANEOUS EQUATIONS
ENTER YOUR EQUATIONS AS FOLLOWS:
10 F(0)=3 * X(0) – 7 * X(1) + 5:RETURN
11 F(1)=2 * X(0) + 3 * X(1) – 1:RETURN

FOR DIFFERENTIAL EQUATIONS
ENTER EQUATIONS AS FOLLOWS:
FOR 0TH ORDER:- START ON LINE 10
FOR 1ST ORDER:- START ON LINE 11
FOR 2ND ORDER:- START ON LINE 12
FOR 3RD ORDER:- START ON LINE 13
FOR 4TH ORDER:- START ON LINE 14

FOR SIM. DIFF. EQU. USE LINES STARTING
AT 10, 20, 30 & 40 FOR 1, 2, 3 & 4 SETS

PRESS SPACE BAR TO CONTINUE
OR <RESET> TO ENTER EQUATION
```

Since we would like to enter our equations, we would have to press the RESET key. We then type in the problem differential equation first, followed by its derived equations, into lines indicated above (i.e. 12-15). We have already written these equations with their appropriate line numbers in Section 7.3. Remember that we only need to type into the computer the problem equation and its derived equations as the Taylor's series equations are already in the computer.

Re-RUN the program and after pressing the space bar, provide the following initial parameters:

```
DIFFERENTIAL EQUATIONS? (Y/N) Y
SIMULTANEOUS? (Y/N) N

DIFF ORDER IN SUB-SET 1? (0/1/2/3/4) 2
```

```
NR OF EQU IN SUB-SET 1? 4
ENTER INITIAL VALUES
FOR SUB-SET 1
C(0)=1
C(1)=0
H=.1
T=0

CONSTANTS? (Y/N) N
DECIM ACCUR = .000000001
NR OF ITERATIONS 100

OUTPUT RESULTS TO PRINTER? (Y/N) N
RE-ARRANGE EQUATIONS? (Y/N) N
DEFAULT STARTING PARAMETERS ARE:
X()=0, R()=0, SIGNS=-1 (Y/N)? Y
```

at which point the program starts to solve simultaneously the four equations we typed in, together with the two Taylor's series already in lines 10 and 11.

After 17 iterations, the solution for X(2) to X(5) is given in the usual way, but at the very end the following new values of constants are given:

```
C(0)=.9949875
C(1)=-.1005
H=.1
T=.1
```

followed by the message

```
CHANGE H? (Y/N/S/C/Q).
```

Option 'Y' allows us to change the value of H and proceed to the next plotted point, while option 'N' automatically proceeds to the next plotted point without changing H. Option 'S' allows us to re-set H, and re-sequence the equations (this option will be discussed in detail later on), while option 'C' allows us to re-set H and continue the calculation to a specified value of T. All options start from the latest plotted position. Option Q allows us to exit the program.

Returning to our example, we can continue the calculation by responding with C to the question CHANGE H?, causing the computer response:

NEXT LIMIT OF T?

where T is the point at which we wish to stop the calculation. Typing 0.8, we are asked:

NEW VALUE OF H?

in case we wanted to change H for this particular range of T.

Typing 0.1, the computer will continue to the desired point where

C(0)=.600240074
C(1)=-1.32940203.

Changing H to -.1, we can return to the original starting point where

C(0)=1.00161425
C(1)=-2.35583431E-03.

Note that the accumulation of round-off errors has caused us to return to a point which differs by .0016 from the original starting position even though the requested decimal accuracy was very high. Methods for minimizing such errors were discussed in Section 7.5.

8.1 DIFF.ROOTS FORMAT FOR SIMULTANEOUS DIFFERENTIAL EQUATIONS

In order to illustrate the use of the DIFF.ROOTS program with simultaneous differential equations we will return to the bridge cable catenary problem given in Section 7.4, where the following equations were considered:

$$8\,(X'')^2 - 2 - Y\,X' = 0 \text{ and } (Y')^2 - 1 - (X')^2 = 0,$$

together with their respective derived equations

$$16\,X''\,X''' - Y\,X'' - Y'\,X' = 0 \text{ and } 2\,Y'\,Y'' - 2\,X'\,X'' = 0.$$

To format these to the requirements of the DIFF.ROOTS program, we must remember that the Taylor's series equations for each sub-set start on lines 10 and 20, respectively. Since the first equation is of the second order, the actual problem equation must be typed in line 12, while the second problem equation must be typed in line 21. In both cases, the first digit of the line number indicates the sub-set number, while the second digit indicates the differential order of the problem equation. All derived equations for each sub-set are typed in consecutive lines following their respective problem equation. Thus, we obtain:

```
12 F(2)=8*X(2)↑2-2-C(10)*C(1):RETURN
13 F(3)=16*X(2)*X(3)-C(10)*X(2)-X(11)*C(1):RETURN
21 F(11)=X(11)↑2-1-C(1)↑2:RETURN
22 F(12)=2*X(12)*X(11)-2*C(1)*X(2):RETURN
```

LOAD the DIFF.ROOTS program and type in lines 12, 13, 21 and 22. These will automatically overwrite the unwanted Taylor's series equations which normally occupy lines 10-13, 20-23, etc.

On RUNning the program, the operator is asked to define the number of simultaneous problem equations (2 in this case), the differential order of the problem equation in each sub-set (2 and 1 in this case), followed by the number of equations in each sub-set (2 and 2 in this case). Note that the number of equations in each sub-set refers only to the number of equations actually typed in and does not include the Taylor's equations. From that point on, the program works out the correct number of constants (Cs) and allocates to them their correct suffixes prior to requesting their value.

A full program RUN is shown below. After the first plotted point, the 'N' option in the CHANGE H? request is chosen, as opposed to C, in order to observe the solutions step by step.

```
DIFERENTIAL EQUATIONS? (Y/N) Y
SIMULTANEOUS? (Y/N) Y
HOW MANY? (2/3/4) 2
DIFF ORDER IN SUB-SET 1? (0/1/2/3/4) 2
DIFF ORDER IN SUB-SET 2? (0/1/2/3/4) 1
NR OF EQU IN SUB-SET 1? 2
NR OF EQU IN SUB-SET 2? 2
ENTER INITIAL VALUES
FOR SUB-SET 1
C(0)=2
C(1)=0
FOR SUB-SET 2
C(10)=0
H=.1
T=0
```

```
CONSTANTS? (Y/N) N
DECIM ACCUR = .000000001
NR OF ITERATIONS 50

OUTPUT RESULTS TO PRINTER? (Y/N) N
RE-ARRANGE EQUATIONS? (Y/N) N
DEFAULT STARTING PARAMETERS ARE:
X()=0, R()=0, SIGNS=-1 (Y/N)? Y

SIGN COMB. -1 -1 -1 -1
EQU. SEQU. 2  3  11 12
```

ITER	P	R	ROOT
1	1	0	1.44363548
1	1	0	0
1	1	0	.881373587
1	1	0	0
2	1	1	-.246510917
2	2	0	0
2	2	0	1.16028002
2	2	0	0
3	1	2	.0540956502
3	3	0	0
3	2	1	.946295179
3	3	0	0
–	–	–	________
–	–	–	________
–	–	–	________
–	–	–	________
14	10	4	.5
14	14	0	0
14	10	4	1
14	14	0	0

SOLUTION	RESIDUAL
X(2)=.5	F(2)=0
X(3)=0	F(3)=0
X(11)=1	F(11)=0
X(12)=0	F(12)=0

```
SIGN COMB. -1 -1 -1 -1
EQU. SEQU. 2  3  11 12

C(0)=2.0025
C(1)=.0499999996
C(10)=.0999999998
H=.1
T .1

CHANGE H? (Y/N/S/C/Q) N

C(0)=2.01000521
C(1)=.100124961
C(10)=.200249922
H=.1
T=.2

CHANGE H? (Y/N/S/C/Q) N

C(0)= 2.02253438
C(1)=.150500117
C(10)=.301000235
H=.1
T=.3

CHANGE H? (Y/N/S/C/Q) N

C(0)=2.04011881
C(1)=.201251329
C(10)=.402502657
H=.1
T=.4

CHANGE H? (Y/N/S/C/Q) Q
```

WARNING: The DIFF.ROOTS program can only re-sequence equations within any sub-set. Errors will develop should you try to sequence equations between sub-sets. Occasionally when a problem (such as some implicit differential equations) demands equation (problem and its derived equations) sequencing between the sub-sets, then they will have to be retyped into a re-LOADed DIFF.ROOTS in different sub-sets.

Problem:

The following four simultaneous differential equations are to be solved using the DIFF.ROOTS program:

$X''' - Y'' - 2Z' - T^2 = 0$	(3rd order)
$Y'' - Z' - T^2 / 2 = 0$	(2nd order)
$Z' - T^2 / 2 = 0$	(1st order)
$W - 2X + Y - 2Z - 1 = 0$	(0th order)

First differentiate the above problem equations to obtain the requisite number of derived equations and then format the resulting sub-sets into the DIFF.ROOTS requirements.

Solutions:

Problem sub-set 1:

$$X''' - Y'' - 2Z' - T^2 = 0$$

Derived equations:

$$X^{IV} - Y''' - 2Z'' - 2T = 0$$
$$X^{V} - Y^{IV} - 2Z''' - 2 = 0$$

Problem sub-set 2:

$$Y'' - Z' - T^2 / 2 = 0$$

Derived equations:

$$Y''' - Z'' - T = 0$$
$$Y^{IV} - Z''' - 1 = 0$$

Problem sub-set 3:

$$Z' - T^2 / 2 = 0$$

Derived equations:

$$Z'' - T = 0$$
$$Z''' - 1 = 0$$

Problem sub-set 4:

$$W - 2X + Y - 2Z - 1 = 0$$

We can now write these equations in the correct DIFF.ROOTS format, remembering to replace variables X, Y, Z and W with variable X() suitably subscripted. These depend on the respective differential order of each problem equation, the sub-set number and the derivative order of each variable.

Thus, we obtain:

```
13 F(3)=X(3)-X(12)-2*X(21)-T↑2:RETURN
14 F(4)=X(4)-X(13)-2*X(22)-2*T:RETURN
15 F(5)=X(5)-X(14)-2*X(23)-2:RETURN

22 F(12)=X(12)-X(21)-T↑2/2:RETURN
23 F(13)=X(13)-X(22)-T:RETURN
24 F(14)=X(14)-X(23)-1:RETURN

31 F(21)=X(21)-T  2/2:RETURN
32 F(22)=X(22)-T:RETURN
33 F(23)=X(23)-1:RETURN

40 F(30)=X(30)-2*X(0) + X(10)-2*X(20)-1:RETURN
```

Typing the above equations into the computer and providing the following initial parameters:

```
C(0)=.05            C(1)=.25      C(2)=1
C(10)=.08333        C(11)=.333
C(20)=.1666
H=.2
T=1
```

and the DEFAULT options for all other parameters, we obtain after the first step:

```
C(0)=.124416        C(1)=.51840       C(2)=1.7280
C(10)=.17280        C(11)=.5760
C(20)=.28801
X(30)=1.652032
```

Note that as the last equation is in fact a substitution equation of zeroth order, there is no C answer associated with it, but we must refer to the value of X(30) for the answer.

These results and further answers can be checked because the primitive equations corresponding to the problem equations are known and given by:

$X = T^5/20$, $Y = T^4/12$, $Z = T^3/6$ and $W = 2X-Y+2Z+1$.

8.2 OTHER DIFF.ROOTS OPTIONS

Occasionally, the solution sequence of differential equations might have to be changed from the normal sequence from which the derived equations were obtained for a given problem equation. The DIFF.ROOTS program allows you to change the equation sequence within any sub-set, either before starting to solve for the first step, or after pressing the space bar, provided H=0, by typing S in answer to the question:

CHANGE H? (Y/N/S/C/Q).

Warning: If the space bar is pressed when H has a value other than 0, the initial parameters (constants) will be lost.

As an example of equations that need sequencing, we will solve the following first order differential equation:

$X X' - 2 T^3 = 0$,

which has the primitive equation $X^2 = T^4$. The derived equations are:

$X X'' + (X')^2 - 6 T^2 = 0$
$X X''' + 3 X' X'' - 12 T = 0$
$X X^{IV} + 4 X' X''' + 3 (X'')^2 - 12 = 0.$

Typing these into the computer in the following DIFF.ROOTS format:

```
11 F(1)=C(0)*X(1) - 2*T↑3:RETURN
12 F(2)=C(0)*X(2) + X(1)↑2-6*T↑2:RETURN
13 F(3)=C(0)*X(3) + 3*X(1)*X(2)-12*T:RETURN
14 F(4)=C(0)*X(4) + 4*X(1)*X(3) + 3*X(2)↑2-12:RETURN
```

and running the program with the initial starting conditions:

C(0)=0
H=0
T=0

and the DEFAULT options for all other parameters, we obtain divergent results. On seeing the answers diverge, press the space bar to obtain the response

CHANGE H? (Y/N/S/C/Q)

from the computer.

Now type S, in which case you will be given the option to first change H and then sequence the equations. Choose a new sequence until, with H=0, the results converge. In this case the sequence resulting in convergence is:

1 4 3 2

with all other parameters being the DEFAULT. On reaching the result

X(2)=2

change H to 0.2 keeping the same equation sequence as above. The first plotted point results are then obtained as:

C(0)=0.04
H=0.2
T=0.2.

In order to continue to the next plotted point,type S, re-set H to 0 (to check convergency) and re-sequence the equations to their original entry sequence, which is virtually always the case, and carry on from that point by changing H to 0.2. Further plotted points are reached by subsequently choosing option C and providing the upper limit of T for which solutions are desired.

8.3 REDUCING THE BURDEN OF DIFFERENTIATION

One criticism levelled at the Taylor's series method for solving differential equations is that occasionally the functions are so complicated that differentiation is a laborious task. We shall first look at simple techniques for reducing such a burden, prior to discussing a method which requires a slight formatting procedure.

Let us assume that we had to differentiate the following function:

$$S' = \frac{dS'}{dT} = \left(1 + \frac{S^2}{4}\right)^{½}$$

Obviously, S′ can be squared to give the easily differentiable function

$$(S')^2 - 1 - \frac{S^2}{4} = 0.$$

Alternatively, we could differentiate the original function to obtain:

$$S'' = \frac{SS'}{4\left(1 + \frac{S^2}{4}\right)^{½}}$$

which, on substituting S′ for the denominator bracket, results in

$$S'' = \frac{S}{4}.$$

This is a much easier function to differentiate repeatedly.

As an example of a formatting procedure, in order to allow the computer to differentiate complicated functions, consider the second order differential equation given below:

$F_2 = F(X'',X',X,T) = F(x_2, x_1, x_0, T) = 0$

Differentiating, we obtain:

$$F_3 = \frac{dF_2}{dT} = \left(\frac{\partial F_2}{\partial X''}\right)\frac{dX''}{dT} + \left(\frac{\partial F_2}{\partial X'}\right)\frac{dX'}{dT} + \left(\frac{\partial F_2}{\partial X}\right)\frac{dX}{dT} + \left(\frac{\partial F_2}{\partial T}\right)$$

$$= \left(\frac{\partial F_2}{\partial X''}\right) x_3 + \left(\frac{\partial F_2}{\partial X'}\right) x_2 + \left(\frac{\partial F_2}{\partial X}\right) x_1 + \left(\frac{\partial F_2}{\partial T}\right) = 0.$$

If the first term is too difficult to differentiate, we can replace it by

$$\frac{1}{2G}\left\{ F(x_2 + G) - F(x_2 - G) \right\} x_3.$$

As an example of the above method, consider the following problem:

$$F_2 = X^2 + TX' + \log \left\{ \cos^{-1} (X'' T^2) \right\}$$

As the first two terms of the expression can be differentiated easily, we will only use the difference formula on the log part to obtain:

$$F_3 = 2XX' + X' + TX'' + \frac{X'''}{2G}\left\{ \log\left[\cos^{-1}(X''+G)T^2 \right] - \log\left[\cos^{-1}(X''-G)T^2 \right] \right\}$$

$$+ \frac{1}{2G}\left\{ \log\left[\cos^{-1}(X'' (T+G)^2) \right] - \log\left[\cos^{-1} (X'' (T-G)^2) \right] \right\}.$$

Both expressions can now be formatted to the DIFF.ROOTS requirements by writing F(2), F(3), C(0), C(1), X(2) and X(3) for F_2, F_3, X, X', X'', and X''', respectively. The $\cos^{-1}$ function can be expressed in terms of BASIC functions as shown in Table A.2 of Appendix A.

You will generally find that overall computational accuracy is unaffected by difference differentiation.

8.4 OSCILLATING COUPLED BODIES

Fig. 8.1 shows two bodies of mass m_0 and m coupled together by two springs having stiffness constants k_0 and k_1. The assembly is attached to the ceiling and can move freely in the vertical direction. Displacements x_0 and x_1 are measured from the shown equilibrium positions of the bodies.

The motion of the coupled system is described by the following two simultaneous second-order differential equations:

$$m_0 \, x_0'' = k_0 \, x_0 - k_1 \, (x_1 - x_0)$$
$$m_1 \, x_1'' = -k_1 \, (x_1 - x_0)$$

Assuming that $m_0 = m_1 = 1$ unit and the stiffness constants of the springs, k_0 and k_1 are equal to 3 and 2 units respectively, then we can proceed to solve the equations, given the initial conditions:

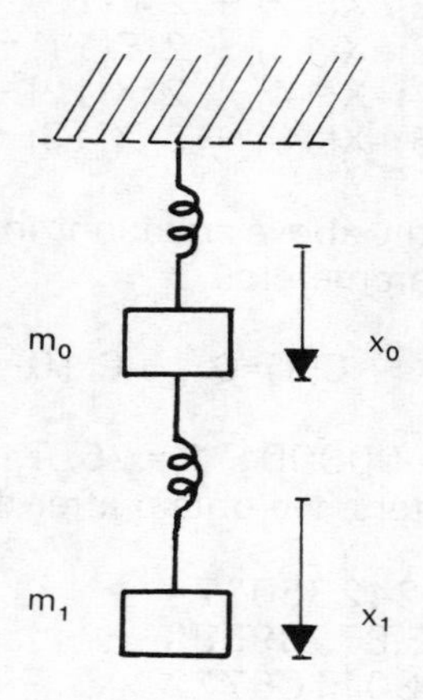

Figure 8.1 Oscillating bodies.

$x_0 = 5, \quad x_0' = 0, \quad x_1 = 0 \quad$ and $\quad x_1' = 0,$

which would correspond to the situation of moving m_0 five units towards m_1, without displacing m_1 from its equilibrium position.

Substituting the above conditions to the two problem equations and their derived functions, we obtain:

$$x_0'' + 5x_0 - 2x_1 = 0 \qquad x_1'' + 2x_1 - 2x_0 = 0$$

$$x_0''' + 5x_0' - 2x_1' = 0 \qquad x_1''' + 2x_1 - 2x_0' = 0$$

$$x_0^{IV} + 5x_0'' - 2x_1'' = 0 \qquad x_1^{IV} + 2x_1'' - 2x_0'' = 0$$

$$x_0^{V} + 5x_0''' - 2x_1''' = 0 \qquad x_1^{V} + 2x_1''' - 2x_0''' = 0.$$

Converting these equations into the DIFF.ROOTS format, remembering that $x_0 = C(0)$, $x_0' = C(1)$, $x_1 = C(10)$ and $x_1' = C(11)$, we obtain:

```
12 F(2)=X(2) + 5*C(0) - 2*C(10):RETURN
13 F(3)=X(3) + 5*C(1) - 2*C(11):RETURN
14 F(4)=X(4) + 5*X(2) - 2*X(12):RETURN
15 F(5)=X(5) + 5*X(3) - 2*X(13):RETURN
```

```
22 F(12)=X(12) + 2*C(10) - 2*C(0):RETURN
23 F(13)=X(13) + 2*C(11) - 2*C(1):RETURN
24 F(14)=X(14) + 2*X(12) - 2*X(2):RETURN
25 F(15)=X(15) + 2*X(13) - 2*X(3):RETURN
```

Typing the above equations into the computer program and providing the initial parameters:

C(0)=5 C(1)=0 C(10)=0 C(11)=0,

with D=.000001, H=.25, T=0 and the DEFAULT options for all other parameters, we obtain after the first step:

C(0)=4.24235027
C(1)=−5.87239599
C(10)=.301106772
C(11)=2.31770835.

Choosing option C (in order to continue), and setting the upper limit for T to 2.75 with the same H as before, we obtain the results shown in Table 8.1.

TABLE 8.1 Displacements of Oscillating Bodies.

T	x_0 = C(0)	x_1 = C(10)
0.25	4.24235027	0.301106772
0.50	2.23642614	1.07574998
0.75	−0.316764977	1.98762145
1.00	−2.53417286	2.61787229
1.25	−3.66856992	2.62263579
1.50	−3.3761721	1.86498837
1.75	−1.83710703	0.47300958
2.00	0.31469596	−1.19763648
2.25	2.22640577	−2.68355716
2.50	3.14037491	−3.57297296
2.75	2.67309966	−3.64724653

These results can be checked because the primitive equations corresponding to the above problem equations are known and given by:

$x_0 = C(0) = 4\cos(\sqrt{6}\,T) + \cos(T)$ and
$x_1 = C(10) = -2\cos(\sqrt{6}\,T) + 2\cos(T)$.

Substituting T=2.75 gives $x_0 = 2.67240518$ and $x_1 = -3.6469585$, which is a good comparison for such a coarse value of H.

Observe that plotting the results will also give the varying crossover period points. As the maximum number of derived equations have been used, greater accuracy can only be obtained with the given number of Taylor's series equations, by decreasing the plotting increment H.

CHAPTER 9

BOUNDARY, EIGEN AND PARTIAL DIFFERENTIAL EQUATIONS

9.1 THE FINITE-DIFFERENCE EQUATIONS METHOD

Whereas a differential equation involves functions defined on some interval of real numbers and their derivatives, a difference equation involves functions defined on some interval of integers and their differences. In earlier sections we considered a numerical method for which a particular solution is determined from the differential equation and all initial information given at a single point. These are initial-value problems.

Problems in which particular solutions are determined from initial conditions, such as slope and function value, given at two or more points, are called boundary-value problems. These equations, which could be heavily non-linear and implicit, are converted to difference equations which are then solved simultaneously with a technique similar to that shown in solving Partial Differential Equations (Section 9.4).

Boundary-value equations normally constitute one of the most difficult class of problems to solve on a computer, especially in time and formatting effort. At the expense of less accuracy, enlightenment will be greatly enhanced if we restrict ourselves to the shorter method of Finite-Difference. By this scheme, every derivative appearing in the differential equation as well as in the boundary conditions, is replaced by an appropriate Lagrangian difference formula approximation.

Differential operators such as X', X'', X''' in differential equations can be most accurately approximated by equal-interval central difference Lagrangian forms to produce a recurrence relationship. If, for reasons that will become shortly apparent, central differences will not fit exactly all the consequent linear finite-difference equations, then the odd remaining equations can be made up by a forward or backward difference Lagrangian forms. Table 9.1, contains first some precise central-difference formulae followed by examples of forward and backward-difference formulae. More precise versions of these equations can be found in relevant texts (see Watson, 1974), but they require more effort in formatting.

Initial-value problems may also be solved by the foregoing method which involves the simultaneous solution of many linear or non-linear finite difference equations in the form of a recurrence relationship incorporating all initial boundary conditions.

The distinct advantage of solving differential equations using finite-difference as opposed to using a Taylor's series initial-value stepping arrangement is that a single run of the SIM.ROOTS program, applied to all the finite-difference equations at once, yields every solution X_i, Y_i, Z_i, at each corresponding value T on the T-axis. Taylor's series method is not directly able to solve boundary-value problems except by the 'shooting' scheme to be touched on later.

TABLE 9.1 Lagrangian Difference Forms

(1) $X'_i \approx (-\tfrac{1}{2}X_{i-1} + \tfrac{1}{2}X_{i+1}) / H$

(2) $\approx (-\tfrac{3}{2}X_{i-1} + 2X_i - \tfrac{1}{2}X_{i+1}) / H$

(3) $X''_i \approx (X_{i-1} - 2X_i + X_{i+1}) / H^2$

Other difference formulae:

(4) $X'_i \approx (-X_i + X_{i+1}) / H$

(5) $\approx (-X_{i-1} + X_i) / H$

(6) $\approx (-\tfrac{3}{2}X_i + 2X_{i+1} - \tfrac{1}{2}X_{i+2}) / H$

(7) $\approx (-\tfrac{3}{2}X_{i-2} + 2X_{i-1} - \tfrac{1}{2}X_i) / H$

(8) $X''_i \approx (X_i - 2X_{i+1} + X_{i+2}) / H^2$

(9) $\approx (X_{i-2} - 2X_{i-1} + X_i) / H^2$

However, the disadvantages of the finite difference formulation could include the need for a larger set of longer simultaneous equations requiring stringent formatting for the same precision as given by Taylor's series method. Generally, a compromise can be reached which can lead to excellent results, even with very difficult boundary-value differential equations.

In order to illustrate the procedure, we will consider a set of simultaneous boundary-value differential equations, whose functions correspond to:

$$X_i = X\,(To + iH),$$

$$Y_i = Y\,(To + iH).$$

In choosing a mesh size H, we divide the range [a,b] (see Fig. 9.1), into n equal intervals of width H = (b-a)/n and we let $T_i = a + iH$ which are the interior mesh points, with i = 0, 1, 2, 3, ..., n.

Sometimes we have to deal with points outside the interval [a,b] called the exterior mesh of fictitious points; those to the left of T_0 being denoted by T_{-1}, T_{-2}, and those to the right of T_n by T_{n+1}, T_{n+2} with corresponding X, Y function subscripts.

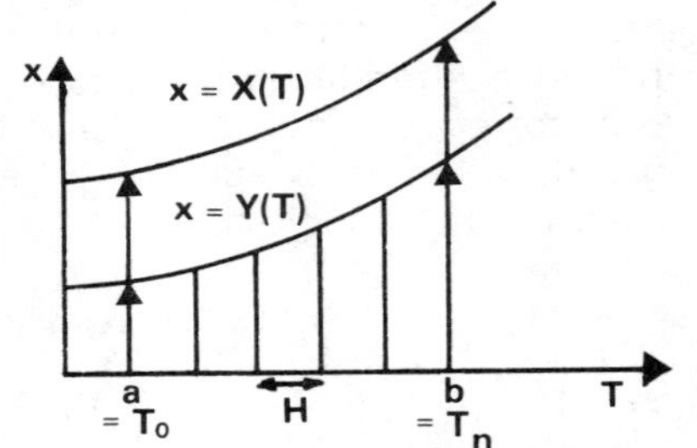

Figure 9.1 Boundary-value problems.

Problem:

Solve the following two simultaneous boundary-value equations:

$$X'' + Y' + 2\sin(T) = 0, \text{ and}$$
$$Y' + X - 4\cos(T) = 0,$$

over the range T = [0, 0.3]. Assume three intervals of mesh size H = 0.1, with boundary conditions:

$$T_0 = 0,\ X_0 = 2,\ Y_0 = 1, \text{ and}$$
$$T_3 = 0.3,\ X_3 = 2.20619319$$

Solution:

Substituting the central difference Lagrangian formulae (1 and 3) into the two problem equations, we obtain:

$$(X_{i-1} - 2X_i + X_{i+1})/H^2 + (-½Y_{i-1} + ½\,Y_{i+1})/H + 2\sin(T_i) = 0$$

$$X_i + (-½Y_{i-1} + ½Y_{i+1})/H - 4\cos(T_i) = 0,$$

which are the recurrence equations. Setting i = 0, 1, 2 and 3, we obtain the following finite difference equations:

$$X_{-1} - 2X_0 + X_1 - ½HY_{-1} + ½HY_1 + 2H^2 \sin(T_0) = 0$$

$$X_0 - 2X_1 + X_2 - ½HY_0 + ½HY_2 + 2H^2 \sin(T_1) = 0$$

$$X_1 - 2X_2 + X_3 - ½HY_1 + ½HY_3 + 2H^2 \sin(T_2) = 0$$

$$X_2 - 2X_3 + X_4 - ½HY_2 + ½HY_4 + 2H^2 \sin(T_3) = 0 \qquad ...(9.1)$$

$$HX_0 - ½Y_{-1} + ½Y_1 - 4H\cos(T_0) = 0$$

$$HX_1 - ½Y_0 + ½Y_2 - 4H\cos(T_1) = 0$$

$$HX_2 - ½Y_1 + ½Y_3 - 4H\cos(T_2) = 0$$

$$HX_3 - ½Y_2 + ½Y_4 - 4H\cos(T_3) = 0$$

The first and last equation of both sets are not included as they contain terms in X and Y which lie outside the range under consideration. This leaves us with only four simultaneous equations, but five unknowns.

A fifth equation can be obtained by substituting the forward difference Lagrangian formula (6 in Table 9.1) into the second recurrence equation to give:

$$HX_0 - \frac{3}{2}Y_0 + 2Y_1 - \frac{1}{2}Y_2 - 4H\cos(T_0) = 0$$

which we will insert between the two sets of Equations (9.1).

Substituting the initial boundary conditions:

$X_0 = 2$, $Y_0 = 1$, at $T_0 = 0$, and
$X_3 = 2.20619319$, at $T_3 = 0.3$,

and writing X(0), X(1), X(2), X(3) and X(4) for X_1, X_2, Y_1, Y_2, and Y_3, we can format the five equations to the SIM. ROOTS requirements as follows:

```
10 F(0)=X(1)-2*X(0) + .05*X(3) + 1.95199667: RETURN
11 F(1)=X(0)-2*X(1)-.05*X(2)+.05*X(4)+2.21016657:RETURN
12 F(2)= 2*X(2)-.5*X(3)-1.7:RETURN
13 F(3)=.1*X(0) + .5*X(3)-.898001666:RETURN
14 F(4)=.1*X(1) -.5*X(2) + .5*X(4)-.392026631:RETURN
```

Running the program and setting all initial Xs and Rs equal to 0, with SIGNs similar to those of the constants above, i.e. +1 +1-1-1 -1, we obtain:

```
X(0)=2.08985918
X(1)=2.1588201
X(2)=1.19450787
X(3)=1.3780315
X(4)=1.54679711
```

Converting back to the original equation notation and comparing the results with those obtained from the analytic solution of the two primitive equations:

$$X = 2\cos(T) + \sin(T)$$
$$Y = \cos(T) + 2\sin(T),$$

we obtain:

T	SIM.ROOTS SOLUTIONS	ANALYTIC SOLUTIONS
0		2
0.1	X=X(0) = 2.08986	2.08984
0.2	X=X(1) = 2.15882	2.15880
0.3		2.20619319
0		1
0.1	Y=X(2) = 1.19450	1.19467
0.2	Y=X(3) = 1.37803	1.37740
0.3	Y=X(4) = 1.54679	1.54637

In order to explain how boundary-value problems can be formatted and solved, we have selected simple equations over very few intervals. Such limitations,. however, do not apply normally.

The ultimate accuracy in solving such problems depends upon (a) the fineness of the mesh interval H, (b) the order of the finite difference approximations, and (c) the sophistication of the chosen Lagrangian forms. As the mesh size is made finer for the same range, the number of equations to be solved increases, as does the formatting effort and program execution time.

Similarly, although the use of higher-order approximations for the same mesh size can increase accuracy, it does so at considerable increase in complexity particulary in regions away from the boundary conditions.

Inaccuracies could result from the presence of a singularity just outside the range of interest in a particular function, especially when using exterior mesh points. Such inaccuracies persist despite attempts at refining the value of H and can only be avoided by changing the form of the differential equation variable(s) before attempting a numerical solution.

Seemingly difficult boundary conditions can exist, such as the case of a function for which $X \rightarrow 0$ as $T \rightarrow \infty$. In such a case, we can approximate and cut off the function at a chosen point T_n, at which we set $X_n = 0$. We can then solve for X_i at points T_i for a given mesh H. The calculation is then repeated with the same mesh size, but this time we increase the cut off position of the function to T_{n+1} at which point we set $X_{n+1} = 0$. Increasing the number of corresponding equations, we proceed to solve for X_i at T_i. If the newly obtained solutions are not within the prescribed accuracy, we increase the cut off point of the function and keep on doing so, until the desired accuracy is reached.

Another possible initial boundary condition could be the slope of the function (X'_0) at the end point T_0, rather than the magnitude of the function (X_0) at the same point. As an example consider the previous problem (Equations 9.1) with the new, typical, boundary condition

$$X'(T_0) + \mu X(T_0) = 0,$$

where μ is a given constant.

Replacing the first term of the above expression by the Lagrangian forward difference formula (4) (see Table 9.1), we obtain:

$$(-X(T_0) + X(T_0+H)/H + \mu X(T_0) = 0,$$

where $X(T_0) = X_0$ and $X(T_0 + H) = X_1$.

Simplifying the above expression, we get:

$$X_0 = X_1 / (1 - \mu H).$$

Thus, if in the set of finite difference equations (Equations 9.1), we replace the unknown X_0 by the expression $X_1/(1-\mu H)$, all other terms remaining unchanged, the resulting system of equations can be solved as before.

9.2 THE SHOOTING METHOD

This scheme solves second order differential equations twice by starting from, say, the left boundary condition (X_0, T_0) using a stepwise initial-value Taylor's series method and progressing to the other boundary at T_n.

The solution starts by first assuming two initial slope values which we will denote (see Fig. 9.2), by:

${}_1X'_0$ and ${}_2X'_0$.

one for each set of otherwise identical differential equations.

Let us suppose that as a result of stepping along the T-axis we obtain corresponding solutions at T_n of ${}_1X_n$ and ${}_2X_n$. If either of these solutions are sufficiently close to the boundary solution X_n then we would have the desired curve. If, however, neither ${}_1X_n$ or ${}_2X_n$ are near enough to the boundary condition, then we must use linear interpolation to deduce a more satisfactory trial value for X'_0 (i.e. ${}_3X'_0$), given by:

$${}_3X'_0 = {}_1X'_0 + \frac{[{}_2X'_0 - {}_1X'_0]\,[X_n - {}_1X_n]}{[{}_2X_n - {}_1X_n]}$$

Using this value of X'_0 as a new initial condition, "shoot" for a closer X_n, and so on.

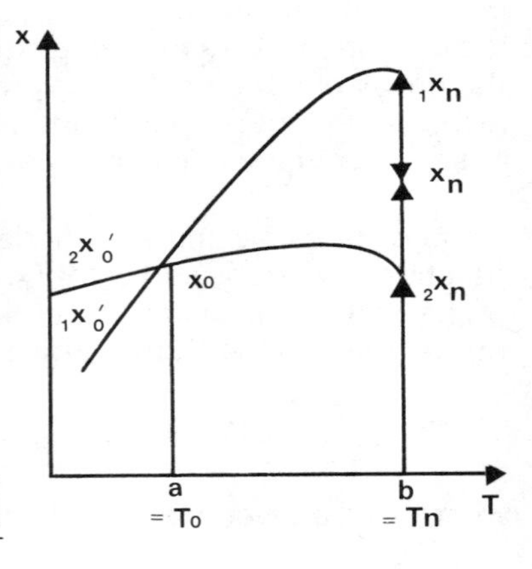

Figure 9.2 Shooting method.

9.3 EIGENVALUE PROBLEMS

Eigenvalues, commonly denoted by λ, arise in problems dealing with oscillating systems, such as the values of λ in the following equation:

$$y'' + \lambda\, q(T)\, y = 0 \qquad (9.2)$$

with boundary conditions y=0 at T=0 and T=b. The function q(T) is dependent on the particular problem under consideration. If q(T)= 1, Equation 9.2 reduces to a homogeneous differential equation, whose analytic solution leads to eigenvalues given by

$$\lambda = (n\pi/b)^2 \text{ for } n = 1, 2, 3 \ldots$$

with corresponding eigenfunctions given by $A\sin(n\pi/b)$, where A is an arbitray constant.

The case of $\lambda = 0$ at n=0 (known as the trivial solution) is excluded as the eigenfunction is identically zero. In the case of a vibration problem, the eigenvalues determine the natural frequencies of the system for which y=0.

Alternatively, equations of motion for N oscillating bodies coupled together by strings or springs, will give a set of N simultaneous homogeneous differential equations. As an example, we examine the case of three such bodies, each of mass m, attached to springs of equal length l as shown in Fig. 9.3. If the two unattached spring ends are fixed and the bodies are under transverse vibration, then the equations of motion for the three bodies are given by:

$$my_1'' = -ky_1/l + k(y_2 - y_1)/l = 0$$

$$my_2'' = -k(y_2 - y_1)/l + k(y_3 - Y_2)/l = 0$$

$$my_3'' = -k(y_3 - y_2)/l - ky_3/l = 0, \qquad (9.3)$$

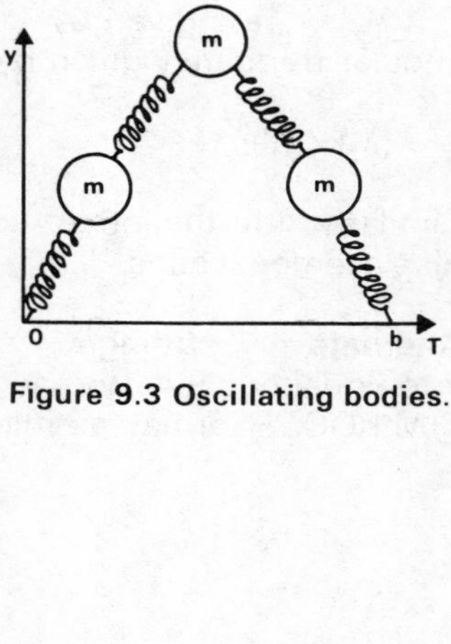

Figure 9.3 Oscillating bodies.

where y_1, y_2 and y_3 are the displacements of the three bodies and k is the spring constant.

If all displacements vary sinusoidally with time, then we can write

$$y_i = Y_i e^{j\omega t}, \ (i=1, 2, 3)$$

which after substitution into Equations 9.3 gives the set of simultaneous linear equations:

$$\begin{aligned}(2-\lambda)\,Y_1 \quad -Y_2 \qquad\qquad &= 0\\ -Y_1 + (2-\lambda)\,Y_2 \qquad -Y_3 &= 0\\ -Y_2 + (2-\lambda)\,Y_3 &= 0,\end{aligned} \tag{9.4}$$

where $\lambda = \omega^2 ml/k$.

These three equations are identical in form to the boundary-value equations resulting from substitution of the central difference formula (3) of Table 9.1 into Equation 9.2. Equations 9.4 have non-trivial solutions when their determinant $\Delta = 0$. This condition yields a polynomial in λ (in this case a cubic), given by:

$$(2-\lambda)^3 - 2(2-\lambda) = 0.$$

The above polynomial can be solved using the ROOTS program to give:

$\lambda_1 = 0.5857864$, $\lambda_2 = 2$ and $\lambda_3 = 3.4142135$.

Substituting these values separately back into Equations 9.4, yields the corresponding displacements:

$Y_1 = (1)\,C$, $Y_2 = (1.4142136)\,C$ and $Y_3 = (1)\,C$,

where C is an arbitrary constant and the bracketed quantities are the eigenvectors. The mode of oscillation is such that the ratios

$Y_1 : Y_2 : Y_3$ are given by 1 : 1.4142136 : 1, and this takes place at an angular frequency given by:

$$\omega = (\lambda\, k/ml)^{1/2}.$$

Similarly, for the other two values of λ we can find corresponding displacement ratios.

Alternately, putting $\lambda = X_0$, $Y_2 = X_1 Y_1$ and $Y_3 = X_2 Y_1$ and substituting into Equations 9.4, we can write the three simultaneous equations in the SIM.ROOTS format, as follows:

```
10 F(0) = ( 2 - X(0) ) - X(1):RETURN
11 F(1) = -1 + ( 2 - X(0) ) * X(1) - X(2):RETURN
12 F(2) = - X(1) + ( 2 - X(0) ) * X(2):RETURN
```

which can be solved directly with all Xs=0, SIGNs=1 and the indicated equation sequence, to obtain:

Equation sequence		2,0,1	0,1,2	0,2,1
λ = X(0)	=	.5857864	3.4142135	2
Y_2 / Y_1 = X(1)	=	1.4142136	−1.4142135	0
Y_3 / Y_1 = X(2)	=	1	1	−1

The ability to solve such problems is vital in explaining vibration modes in rods, strings, organ pipes, cavity resonators, car suspensions and waves of all types.

9.4 PARTIAL DIFFERENTIAL EQUATIONS

The following two-dimensional example of Laplace's equation is an elliptical partial differential equation applied to the distribution of temperature over the surface of a hot metal plate due to heat flow in the plate. Laplace's equation presupposes no sources or sinks of heat over the plane of the plate except at the boundaries.

The usual method for solving partial differential equations is to replace the partial derivatives $\partial T/\partial y$, $\partial^2 T/\partial z^2$ etc by difference quotients, resulting in a difference equation corresponding to each point at the intersections (nodes) of an equal mesh grid (parallel to the axes, as shown in Fig. 9.4) that subdivides the region in which the function values are to be found. Solving these simultaneous equations results in function values at each node which approximate the true values.

Laplace's equation states that:

$$\nabla^2 T(y_i, z_j) = \frac{\partial^2 T}{\partial y^2} + \frac{\partial^2 T}{\partial y^2} = 0,$$

where T is the temperature.

Replacing the derivatives by difference quotients which approximate the derivatives at the point (y_i, z_j), we get, after letting the mesh size $\triangle y = h$ and $\triangle z = k$, the following equation:

$$\triangle^2 T(y_i,z_j) = \frac{1}{h^2}\,[\,T(y_{i+1}, z_j) - 2T(y_i, z_j) + T(y_{i-1}, z_j)] +$$

$$\frac{1}{k^2}\,[\,T(y_i, z_{j+1}) - 2T(y_i, z_j) + T(y_i, z_{j-1})] = 0.$$

It is convenient to let double subscripts on T indicate the y, z values. Allowing h=k, we obtain:

$$\nabla^2 T_{ij} = \frac{1}{h^2}[\,T_{i+1,j} + T_{i-1,j} + T_{i,j+1} - 4T_{ij}] = 0.$$

The five points involved in this relationship show a point each above, below, left and right of the central point (y_i,z_j). Hence, we obtain the pictorial operator so fundamental to work in elliptic partial differential equations:

$$\nabla^2 T_{ij} = \frac{1}{h^2}\left\{\begin{matrix} & 1 & \\ 1 & -4 & 1 \\ & 1 & \end{matrix}\right\} T_{ij} = 0.$$

Provided T is sufficiently smooth in the (y,z) plane, the above approximation has an error of order h^2.

We shall illustrate the technique by finding the steady state temperatures on a square metal plate with a 20 cm side. The top and right edges are held at 100º C, while the remaining two edges are held at 0º C. It should be noted that neither the thermal properties of the metal, its size, nor the ambiguous corner temperatures where 100º C and 0º C intersect enter into the calculations.

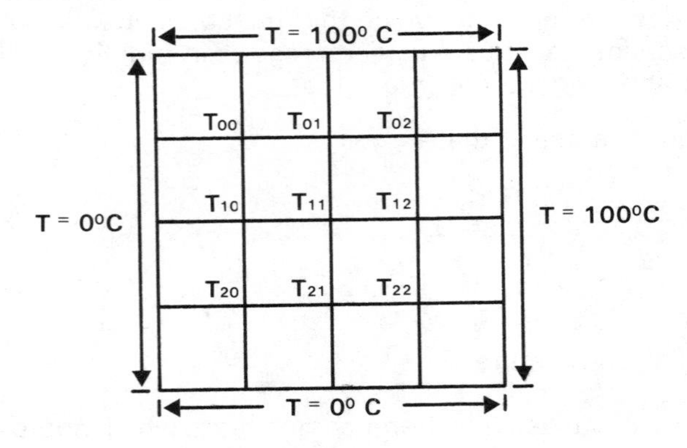

Figure 9.4 Temperature distribution in a square metal plate.

Subdividing the plate into nodes of 5 cm (i.e. h=5) and keeping a clockwise pattern of writing T_{ij}, we obtain the 9 linear equations for i=0, 1, 2; j=0, 1, 2, as shown below:

$$0 + 100 + T_{01} + T_{10} - 4T_{00} = 0$$

$$T_{00} + 100 + T_{02} + T_{11} - 4T_{01} = 0$$

$$T_{01} + 100 + 100 + T_{12} - 4T_{02} = 0$$

$$0 + T_{00} + T_{11} + T_{20} - 4T_{10} = 0$$

$$T_{10} + T_{01} + T_{12} + T_{21} - 4T_{11} = 0$$

$$T_{11} + T_{02} + 100 + T_{22} - 4T_{12} = 0$$

$$0 + T_{10} + T_{21} + 0 - 4T_{20} = 0$$

$$T_{20} + T_{11} + T_{22} + 0 - 4T_{21} = 0$$

$$T_{21} + T_{12} + 100 + 0 - 4T_{22} = 0.$$

We can format the above equations to the SIM.ROOTS requirement by writing X(0)=T_{00}, X(1)=T_{01}, X(2)=T_{02}, X(3)=T_{10}, X(4)=T_{11}, X(5)=T_{12}, X(6)=T_{20}, X(7)=T_{21} and X(8)=T_{22}, to obtain:

```
10 F(0)=-4*X(0) + X(1) + X(3) + 100:RETURN
11 F(1)=X(0)- 4*X(1) + X(2) + X(4) + 100:RETURN
12 F(2)=X(1)-4*X(2) + X(5) + 200:RETURN
13 F(3)=X(0)- 4*X(3) + X(4) + X(6):RETURN
14 F(4)=X(1) + X(3)-4*X(4) + X(5) + X(7):RETURN
15 F(5)=X(2) + X(4)-4*X(5) + X(8) + 100:RETURN
16 F(6)=X(3) + X(7)- 4*X(6):RETURN
17 F(7)=X(6) + X(4) + X(8)-4*X(7):RETURN
18 F(8)=X(7) + X(5)- 4*X(8) + 100:RETURN
```

RUNning the program and supplying the initial parameters:

```
NR OF EQUATIONS? 9
DECIM ACCUR = .0001
ALL Xs = 0
ALL Rs = 0
ALL SIGNs = 1,
```

we obtain the following results in 37 iterations:

$X(0) = T_{00} = 50.0000$
$X(1) = T_{01} = 71.4286$
$X(2) = T_{02} = 85.7143$
$X(3) = T_{10} = 28.5714$
$X(4) = T_{11} = 50.0000$
$X(5) = T_{12} = 71.4286$
$X(6) = T_{20} = 14.2857$
$X(7) = T_{21} = 28.5714$
$X(8) = T_{22} = 50.0000.$

Finer subdivisions of h will result in greater temperature precision, but at the expense of increased number of equations.

Finally, other types of partial differential equations can be treated in a similar manner to the above.

CHAPTER 10

CONSTRAINTS IN EQUATIONS

Frequently, we need to try out different tendencies of equations or functions, or we need to automatically select or reject certain terms or whole equations from computation. In such cases, we need to apply 'logical constraints' or '0-1 variables' to equations, before setting out to solve them.

Historically, Langrange's method (discussed elsewhere in this book) was the first scheme in which functional constraints could be injected into expressions. In this Chapter, we can do little more than introduce constraints on individual functions and equations.

10.1 NUMERICAL CONSTRAINTS

Often, the time taken in table look-up procedures for functional constraints can be greatly reduced in programs by using numerical constraints. Thus, when certain constraints (i.e. 0, 1, 2 ...) are entered into the computer, certain terms in formulae or equations can be rejected.

As an example, consider the following equation:

$$F_0 = x_0 + x_1 (1-B_3) + x_2 (1-B_2) + x_3 (1-B_1) + x_4 B_3 + x_5 B_2 + x_6 B_1 + x_7 + Dx_8. \qquad (10.1)$$

If, for instance, B_3 is made unity, then obviously the term x_4 will be included as a term in F_0, but x_1 will be excluded by virtue of the fact that $(1-B_3)$ is equal to 0.

Now, we could choose the values of all B_i to satisfy the truth table given below which depends on values of Q within the range 1 to 4.

TABLE 10.1 Single Variable Truth Table

Q = 1 2 3 4
B_1 = 1 1 1 0
B_2 = 1 1 0 0
B_3 = 1 0 0 0

We would like to have a simple formula for each B_i to be incorporated into Equation (10.1), so that, as the value of Q changes, the appropriate value of B_i (1 or 0) is obtained, thus allowing selection or rejection of the corresponding x_j terms in Equation (10.1).

One possible expression for B_1 would be:

$$B_1 = (Q-2)(Q-3)(Q-4)/k_1 + (Q-1)(Q-3)(Q-4)/k_2 +$$

$$(Q-1)(Q-2)(Q-4)/k_3 + 0/k_4, \qquad (10.2)$$

in which the first term is zero when Q=2,3,4, but not when Q=1, whereas both the other terms containing (Q-1) terms are zero when Q=1. However, we need the first term to be equal to unity when Q=1, because B_1 must be equal to 1 from the truth table.

Thus, when Q=1, then

$$B_1 = (1-2)(1-3)(1-4)/k_1 = 1$$

which infers that k_1 must be equal to −6. Similarly, when Q=2 all terms of B_1 are eliminated save the second, from which k_2 can be evaluated as 1/6, while when Q=3, k_3 also evaluates to 1/6.
Finally, as all three terms of Equation (10.2) contain the (Q−4) factor, they are all equal to zero when Q=0.

We can simplify the resultant expression for B_1, to obtain:

$$B_1 = (Q-4)(-Q^2 + 2Q-3)/6$$

which indeed satisfies the truth table for all four values of Q.

Similarly, it can be shown that:

$B_2 = (Q-4)(Q-3)(2Q-1)/6$, and

$B_3 = (Q-3)(Q-4)(2-Q)/6$.

The above expressions for B_1, B_2 and B_3 can either be substituted into Equation (10.1) or calculated separately. As a matter of interest, note that should we neglect the last term of Equation (10.1), we would obtain an expression which is identical in form to that of the Taylor's series expression used on line 10 of the DIFF.ROOTS program (see listing in Appendix F). In that case however, the expression was evaluated by yet another method.

Let us assume now that the constraint D in the last term of Equation (10.1) must satisfy the following bi-variable truth table:

TABLE 10.2 Bi-variable Truth Table

E	Q = 1	2	3	4
0	D = 0	0	0	0
1	D = 0	0	0	1
2	D = 0	0	1	X
3	D = 0	1	X	X

where entry X signifies 'not applicable' or 'arbitrary'.

Whatever the resultant expressions for D turns out to be, it must be multiplied by E(Q-1) because when E=0 or Q=1, the corresponding E row and Q column must all be zero. Note that there are three 1's in the remaining blocked off section of the table, so that there will be three terms in the D formula, the upper one of which, say $D_{1,4}$, can be estimated (in the same manner as in the case of B_i) to be:

$D_{1,4} = E(Q-1)\ [(Q-2)(Q-3)/k_1\ ]$,

which eliminates the Q=2 and Q=3 columns.

The "1" in the next row down, leads to $D_{2,3}$ which is also a simple expression because it does not need to contain the irrelevant terms (E–2) and (Q–4), as they are marked by X in the table.
Hence, in order to eliminate the Q=2 column and the E=1 row, we let

$$D_{2,3} = E(Q-1)\ [(Q-2)(E-1)/k_2].$$

Similarly, in order to reject the E=1,2 rows in calculating $D_{3,2}$, we let

$$D_{3,2} = E(Q-1)\ [(E-1)(E-2)/k_3],$$

We calculate the values of k_1, k_2 and k_3 in a similar manner as we have done previously. For k, substitute E=1, Q=4 in $D_{1,4}$ to give:

$$D_{1,4} = 1(4-1)\ [(4-2)(4-3)/k_1] = 1,$$

which yields k_1 =6. For k_2, substitute E=2, Q=3 in $D_{2,3}$ to give:

$$D_{2,3} = 2(3-1)\ [(3-2)(2-1)/k_2\ ] = 1,$$

which yields k_2 =4. Finally, for k_3, substitute E=3, Q=2 in $D_{3,2}$ to give:

$$D_{3,2} = 3(2-1)\ [(3-1)(3-2)/k_3] = 1,$$

which yields k_3 =6.

The final expression for D in Equation (10.1) will be the sum of $D_{1,4}$, $D_{2,3}$ and $D_{3,2}$, as follows:

$$D = D_{1,4} + D_{2,3} + D_{3,2}$$

$$= E(Q-1)\ [(Q-2)(Q-3)/6 + (Q-2)(E-1)/4 + (E-1)(E-2)/6],$$

which is a single expression that satisfies all thirteen table entries. This form for expressing D is not necessarily unique.

10.2 AN ALGEBRAIC COUNTER ALGORITHM

Consider the algebraic operator

$$A(N) = [2N - 1 + (-1)^N\ ]/4, \tag{10.3}$$

where N is usually an integer, but can be a function such as

$|N|$, 2N+1, N^2 or a further operator such as

$A(A(N-3) = A^2(N-3)$, a double operation (not A squared).

We shall examine the behaviour of Equation (10.3) together with that of another operator, given by

$$C(N) = (1-(-1)^N/2, \qquad (10.4)$$

which has N defined in a similar way to that of Equation (10.3).

Table 10.3 shows some of the result obtained by applying the above two operators to integers N.

TABLE 10.3 A(N) and C(N) Operators Applied to Integers N

N	0	1	2	3	4	5	6	7	8	9	10	11	12
A(N)	0	0	1	1	2	2	3	3	4	4	5	5	6
A^2 (N)	0	0	0	0	1	1	1	1	2	2	2	2	3
A^3 (N)	0	0	0	0	0	0	0	0	1	1	1	1	1
C (N)	0	1	0	1	0	1	0	1	0	1	0	1	0
C(A(N))	0	0	1	1	0	0	1	1	0	0	1	1	0
C(A^2 (N))	0	0	0	0	1	1	1	1	0	0	0	0	1
C(A^3 (N))	0	0	0	0	0	0	0	0	1	1	1	1	1

Observe that the algebraic C operators form a binary sequence. Other codes and number series can also be replicated by the above means. Note that the C(N) operator codes are cyclical, while those resulting from the A(N) operator are not.

The following example illustrates the use of the A(N) code of Table 10.3. We wish to investigate the function

$F = \tan(x) - 2x$

in all quadrants, from N=1 to 6. Thus, function F has the following form:

Quadrant (N)	Function (F)
1	$\tan(x) - 2x$
2	$\tan(x-\pi) - 2x$
3	$\tan(x-\pi) - 2x$
4	$\tan(x-2\pi) - 2x$
5	$\tan(x-2\pi) - 2x$
6	$\tan(x-3\pi) - 2x$

which are precisely represented by

$F = \tan(x-A(N)\pi) - 2x.$

Many other codes can be generated by phase shifting the sequences up and down the N integers. Thus, the B constraints of Table 10.1 can be reproduced as follows:

We require B_1 to take the values of 1, 1, 1, 0 for values of Q equal to 1, 2, 3, 4. First, find the sequence of B_1 in Table 10.3, as indicated below

N	0	1	2	3	4	5	6	7	8	9
$C(A^2(N))$	0	0	0	0	1	1	1	1	0	0

then, substitute Q for N and add 4 to it so that when Q=1 it corresponds to the N=5 position of the above sequence, to give

$B_1 = C(A^2(Q+4)).$

Similarly, for B_2 which takes the values of 1, 1, 0, 0 when Q equals 1, 2, 3, 4, we can find the required sequence in the C(A(N)) operator of Table 10.3, as follows:

N	0	1	2	3	4	5	6
C(A(N))	0	0	1	1	0	0	1

which gives B_2 as:

$B_2 = C(A(Q+1))$.

Finally, B_3 can be obtained, in exactly the same way, as:

$B_3 = C(A^2\,(Q+6))$.

Obviously, the above operators are not unique and only experience will give you the ability to choose simpler ones. For example, the above B_i sequences can be replicated by much simpler operators which are entirely composed of the A(N) operator of Table 10.3, as follows:

$B_1 = 1 - A\,(|Q-2|)$, $B_2 = 1 - A(Q-1)$, and $B_3 = A(|\,Q-3\,|)$.

Similarly, the D_{ij} operators describing the boxed-off sequences of Table 10.2 can be easily expressed in terms of the A^2 (N) operator of Table 10.3, as follows:

$D_{1,4} = A^2\,(Q)$, $D_{2,3} = A^2\,(Q+1)$, and $D_{3,2} = A^2\,(Q+2)$.

To replicate the entire bi-variable truth table given in Table 10.2, we need a three-part function for D in terms of E, such that:

When E=1, only the first part of the function evaluates to unity (the other two parts of it evaluating to zero),
when E=2, only the second part evaluates to unity,
when E=3, only the third part evaluates to unity, and
when E=0, all three parts evaluate to zero.

The type of sequence we are looking for is 0, 1, 0, 0, so that the 1 can be appropriately phase shifted in the three-term operator in order to cancel the unwanted terms. This sequence does not exist in Table 10.3, but can, nevertheless, be synthesized from existing sequences in the Table. For example, substituting E for N in Table 10.3, we obtain:

E =	0	1	2	3	4	5	6	7
A^2 (E) =	0	0	0	0	1	1	1	1

and by shifting the last sequence to the left, first by 3 and then by two, we obtain:

$$A^2\ (E+3) = 0 \quad 1 \quad 1 \quad 1 \quad 1 \quad 2 \quad 2 \quad 2$$

$$A^2\ (E+2) = 0 \quad 0 \quad 1 \quad 1 \quad 1 \quad 1 \quad 2 \quad 2$$

respectively. These can now be subtracted to obtain the required operator when E=1, as follows:

$$A^2\ (E+3) - A^2\ (E+2) = 0 \quad 1 \quad 0 \quad 0 \quad 0 \quad 1 \quad 0 \quad 0$$

with the other two conditions for E being satisfied by:

$$A^2\ (E+2) - A^2\ (E+1) = 0 \quad 0 \quad 1 \quad 0 \quad 0 \quad 0 \quad 1 \quad 0$$

$$A^2\ (E+1) - A^2\ (E) \quad = 0 \quad 0 \quad 0 \quad 1 \quad 0 \quad 0 \quad 0 \quad 1.$$

Thus, the final operator for D is obtained by combining the last three three-term operators with their corresponding D operators given earlier, as follows:

$$D = [\, A^2\ (E+3) - A^2\ (E+2)\,]\ A^2\ (Q) + [\, A^2\ (E+2) - A^2\ (E+1)\,]\ A^2\ (Q+1) + [\, A^2\ (E+1) - A^2\ (E)\,]\ A^2\ (Q+2).$$

10.3 ALGEBRAIC CODE CONVERSION

A different approach is needed when converting from one set of 0-1 data variables to another set of 0–1 constraints. The truth table given in Table 10 4 holds for two inputs (A and B). The equivalent algebraic formulae in the table are described by reference to the truth table and in conjunction with certain rules to be given shortly. For completeness, the logical functions describing these entries are also included in the right-hand side of the table.

Naturally, many more input variables than A, B are permissable and similar tables can be constructed by conforming to the same approach. The method is equally adept to non-binary number sequences.

TABLE 10.4 Algebraic and Logical Expressions Describing the Output Constraints of a Two 0-1 Data Variables

					EQUIVALENT ALGEBRAIC FORMULA	LOGICAL FUNCTION
INPUT A	0	1	0	1		
INPUT B	0	0	1	1		
CONSTRAINTS						
(1)	0	0	0	0	0	0
(2)	0	0	0	1	ab	AND
(3)	0	0	1	0	(1-a)b	$\overline{A}.B$
(4)	0	0	1	1	b	B
(5)	0	1	0	0	a(1-b)	$A.\overline{B}$
(6)	0	1	0	1	a	A
(7)	0	1	1	0	$(a-b)^2$	EX/OR
(8)	0	1	1	1	a+b-ab	OR
(9)	1	0	0	0	(1-a) (1-b)	NOR
(10)	1	0	0	1	$1-(a-b)^2$	EX/NOR
(11)	1	0	1	0	1-a	NOT A
(12)	1	0	1	1	1-a (1-b)	$\overline{A.\overline{B}}$
(13)	1	1	0	0	1-b	NOT B
(14)	1	1	0	1	1-(1-a)b	$\overline{\overline{A}.B}$
(15)	1	1	1	0	1-ab	NAND
(16)	1	1	1	1	1	1

Note that the B_i sequence of Table 10.1 is reproduced by the three algebraic formulae corresponding to the NAND, NOT B, and NOR entries of the above table.

In Table 10.4 and in the rest of this section, we use upper case letters to indicate Boolean variables and lower case letters to indicate the equivalent algebraic 0-1 variables. We further define the Boolean operators AND and OR by a dot and an encircled cross, i.e. A.B and $A \oplus B$, respectively.

Our main objective is to simplify and convert either Boolean expressions or truth tables into algebraic expressions which can be manipulated and easily inserted into a computer program. Although Boolean algebra can be used to simplify and reduce logical expressions, the result is not always in a form suitable for inclusion into a program containing algebraic functions or equations.

Since we have defined, at least for this section, algebraic logic variables as having values of 1 or 0, we can postulate that the identity

$$a^N = a$$

is true for all positive integer values of N. We further express the inverse of a (when input A=0), as:

$$\bar{a} = (1-a),$$

and hence, we can express a variable a, as:

$$a = 1-\bar{a} = 1-(1-a).$$

Thus, from the truth table of Table 10.4 (and bearing in mind the above ground rules), we can write down algebraic expression which can be manipulated by standard algebra to give the reduced entries appearing in the middle of the table against the numbered constraints, in the following manner:

(1) to (3): self-explanatory

(4): $\bar{a}b + ab = (1-a)b + ab = b$

(5): self-explanatory

(6): $a\bar{b} + ab = a(1-b) + ab = a$

(7): $a\bar{b} + \bar{a}b = a(1-b) + (1-a)b = a+b-2ab = (a-b)^2$

(8): $a\bar{b} + \bar{a}b + ab = a(1-b) + (1-a)b + ab = a+b-ab$

(9): self-explanatory

(10): $\bar{a}\bar{b} + ab = (1-a)(1-b) + ab = 1-(a-b)^2$

(11): $\bar{a}\bar{b} + \bar{a}b = (1-a)(1-b) + (1-a)b = 1-a$

(12): $\bar{a}\bar{b} + \bar{a}b + ab = (1-a)(1-b) + (1-a)b + ab = 1-a(1-b)$

(13): $\bar{a}\bar{b} + a\bar{b} = (1-a)(1-b) + a(1-b) = 1-b$

(14): $\bar{a}\bar{b} + a\bar{b} + ab = (1-a)(1-b) + a(1-b) + ab = 1-(1-a)b$

(15): $\bar{a}\bar{b} + a\bar{b} + \bar{a}b = (1-a)(1-b) + a(1-b) + (1-a)b = 1-ab$

(16): $\bar{a}\bar{b} + a\bar{b} + \bar{a}b + ab = (1-a)(1-b) + a(1-b) + (1-a)b + ab \quad 1.$

The method can become an independent check to Boolean reduction. For example,

$A.\bar{A} = a\bar{a} = a(1-a) = a-a^2 = a-a = 0,$

$A \oplus \bar{A} = 1-(1-a)(1-(1-a)) = 1-a(1-a) = 1-a+a^2 = 1,$

$A \oplus A.B = 1-(1-a)(1-ab) = 1-1+a+ab-a^2 b = a = A$ (Redundancy),

$A \oplus B = 1-(1-a)(1-b) = 1-\bar{a}\bar{b} = \overline{\bar{a}\bar{b}} = \overline{\bar{A}.\bar{B}}$ (De Morgan's),

$A.B.C \oplus \bar{A}.\bar{B}.C \oplus \bar{A}.B.\bar{C} = 1-(1-abc)(1-\bar{a}\bar{b}c)(1-\bar{a}b\bar{c})$

$= abc + \bar{a}\bar{b}c + \bar{a}b\bar{c}$ (Zero cross-product).

This last expression shows that all cross-products, such as $(abc)(\bar{a}\,\bar{b}\,c)$ are equal to zero. Thus, we can use this fact to simplify and reduce algebraic expressions, without first multiplying out such products.

In using algebra to reduce logic expressions, always attempt to get the final result in a form which contains the same type of variables, i.e. $\bar{a}$, $\bar{b}$, $\bar{c}$ or their inverse a,b,c, but not a mixture of the two, as insertion into a computer program becomes less realistic.

We can now use the above technique to replicate the bi-variable constraint D of Section 10.1 (Table 10.2), by relating it to the A,B inputs of Table 10.4. Replacing Q and E in Table 10.2 with A,B and Y,Z inputs, we obtain the following table:

TABLE 10.5 Bi-variable Truth Table (see Table 10.2)

	A	0	1	0	1
	B	0	0	1	1
Y	Z				
0	0	0	0	0	0
1	0	0	0	0	1
0	1	0	0	1	X
1	1	0	1	X	X

(X=Not Applicable)

The three "1"s are given by the algebraic expression:

$$D = y\bar{z}ab + \bar{y}z\bar{a}b + yza\bar{b} = y(1-z)ab + (1-y)z(1-a)b + yza(1-b)$$

$$= \underset{1}{yab} - \underset{X}{yzab} + \underset{1}{zb} - \underset{X}{zab} - \underset{X}{yzb} + \underset{1}{yza}$$

which correspond to the "1"s and "X"s of the table as shown under the resultant terms. As we do not require the terms corresponding to the X entries of the table, we can reduce the expression to:

D = yab+zb+yza = YAB+ZB+YZA.

Note that the upper case letters used in the above formula refer only to the inputs in Table 10.5.

10.4 CONSTRAINTS IN CURVE FITTING

In this section we will show how arbitrary functions such as polynomial or transcendental curves and straight lines can be strung together on the x-axis of a plotting program (such as the one discussed in Appendix E.2), in a predetermined manner as a single function, using the algebraic logic developed in the previous sections.

Table 10.6 is similar to Table 10.3 with X substituted for N, but with only the C operators being tabulated. Here, we write C_1 in place of C(A(N)), C_2 in place of $C(A^2(N))$, etc. For the sake of completeness, we reproduce these operators below, as follows:

$A_1 = [2X-1+(-1)^{X}]/4,$ $\qquad$ $C_1 = [1-(-1)^{A_1}]/2,$

$A_2 = [2A-1+(-1)^{A_1}]/4,$ $\qquad$ $C_2 = [1-(-1)^{A_2}]/2,$

$A_3 = [2A-1+(-1)^{A_2}]/4,$ $\qquad$ $C_3 = [1-(-1)^{A_3}]/2,$

$A_4 = [2A-1+(-1)^{A_3}]/4,$ $\qquad$ $C_4 = [1-(-1)^{A_4}]/2.$

In the table below, we show only C_2, C_3 and C_4, as they are directly related through their appropriate A_i to X. C_1 is, in this case, redundant.

TABLE 10.6 Algebraic Operators C as Functions of X

X	0	1	2	3	4	5	6	7	8	9	10	11	12	13	14	15	16	17	18	19
C_2	0	0	0	0	1	1	1	1	0	0	0	0	1	1	1	1	0	0	0	0
C_3	0	0	0	0	0	0	0	0	1	1	1	1	1	1	1	1	0	0	0	0
C_4	0	0	0	0	0	0	0	0	0	0	0	0	0	0	0	0	1	1	1	1

More columns will be needed should we require to extend the x-axis beyond 19. This can be achieved easily by extending the operators listed above.

We shall now use this table to string together two equations, a parabola and a power curve given by:

$f(x) = 2^2 - (x-2)^2$, and

$f(x) = (x-4)^{1.3} / 2$.

A plot of these two functions is shown in Fig. 10.1 with the parabola shown in broken lines. Thus, to reproduce this plot by the computer, we must write an algebraic function containing the appropriate constraints so that the parabola is drawn from x=0 to x=3, while the power curve is drawn from x=4 to x=19.

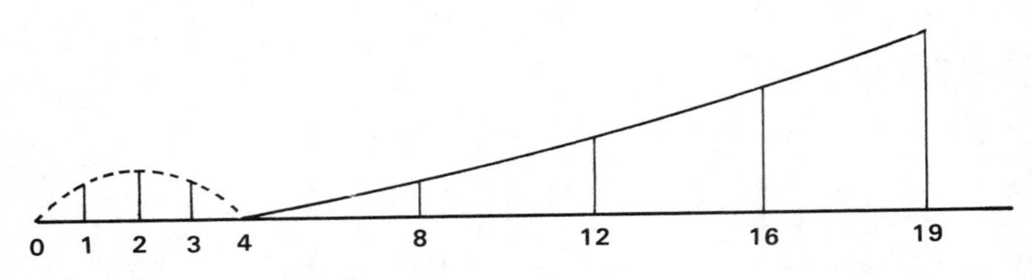

Figure 10.1 A two-function curve

From the figure, we see that the parabola is effective for values of x from 0 to 3 and from Table 10.6, we see that in that range of x all Cs are 0. Therefore, we can express this by:

$$\overline{C}_2 \, \overline{C}_3 \, \overline{C}_4$$

which will have to be multiplied by the function describing the parabola to give:

$$F_1 = \overline{C}_2 \, \overline{C}_3 \, \overline{C}_4 \, (2^2 - (X-2)^2)$$

$$= (1-C_2)(1-C_3)(1-C_4) \, (2^2 - (X-2)^2).$$

The power curve is effective from x=4 to x=19, and is described with common constraints within the four intervals, x=4 to 7, x=8 to 11, x=12 to 15 and x=16 to 19 (see Table 10.6). Thus, we can write:

$$F_2 = \left\{C_2 \, \overline{C}_3 \, \overline{C}_4 + \overline{C}_2 \, C_3 \, \overline{C}_4 + C_2 \, C_3 \, \overline{C}_4 + \overline{C}_2 \, \overline{C}_3 \, C_4\right\}(X-4)^{1\cdot 3} \,/\, 2$$

$$= \left\{(1-C_4) \, [C_2 \, (1-C_3)+(1-C_2)C_3 + C_2 \, C_3 \,] + C_4 \, (1-C_2)(1-C_3)\right\}(X-4)^{1\cdot 3} \,/\, 2.$$

The resultant single function $F = F_1 + F_2$ will reproduce the double curves in the correct sequence and is, therefore, available for further discreet analysis, including differentiation and integration. Other curves could be added to the above function by extending the X and C_i table.

In order to plot the resultant function, LOAD the FUNCTION PLOTTER program (see Appendix E.2) and add the following lines:

```
10 F1=(2*X−1 + (−1)↑X)/4:F2=(2*F1 − 1 + (−1)↑F1)/4:
   C2=(1 − (−1)↑F2)/2:F3=(2*F2 −1 + (−1)↑F2)/4
11 C3=(1−(−1)↑F3)/2:F4=(2*F3 −1 + (−1)↑F3)/4:
   C4=(1−(−1)↑F4)/2
12 F5=(1−C2) * (1−C3) * (1−C4) * (4 − (X−2)↑2)
13 F6=((1−C4) * (C2 * (1−C3) + (1−C2) * C3 + C2 * C3) +
   C4 * (1−C2) * (1−C3)) * ABS((X−4)↑1.3)/2
14 F=F5+F6
```

Note that in line 13 we have used the expression ABS(X-4) ↑ 1.3, because if (X-4) evaluates negative, then raising this to a noninteger power results in a complex value which can occur outside the range of interest of this curve.

On RUNning the program, we obtain the following output:

```
PLOT WIDTH? 40
X-AXIS POSITION? (0 - 37) 20
F-MULTIPLIER? 1
F-SCALE SHIFT? 0
STARTING VALUE OF X? -1
INCREMENT VALUE OF X? 1

!++++! = 5 UNITS
```

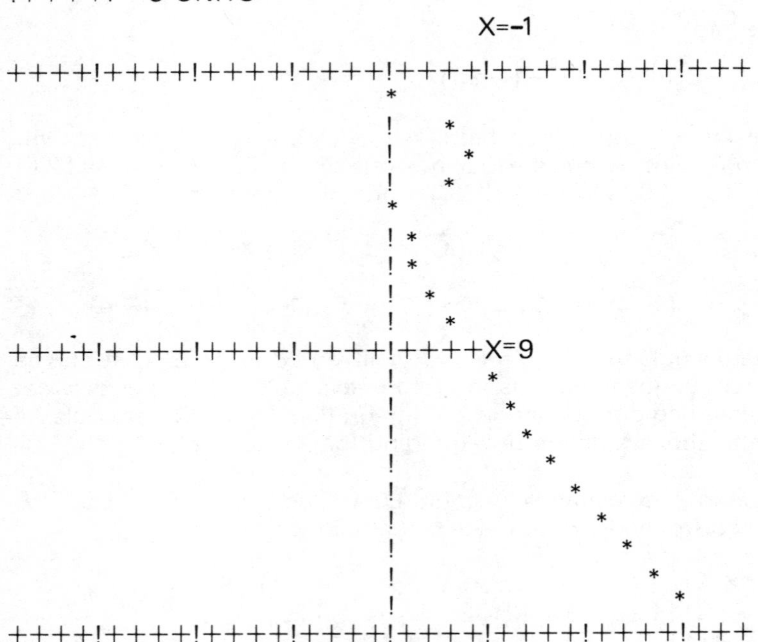

In the above plot, we chose to start at X=-1 in order to show the first point of interest at X=0 as the axis in this program overwrites the plotted points.

APPENDIX A

BASIC MATHEMATICAL FUNCTIONS

Since the equation whose roots you are trying to find may contain standard BASIC functions such as logarithms, square roots, sine of angles, etc., we list below some of the most common functions used in BASIC.

TABLE A.1 Standard BASIC Functions

Call Name	Function
SIN(X)	Sine of angle X, where X is in radians
COS(X)	Cosine of angle X, where X is in radians
TAN(X)	Tangent of angle X, where X is in radians
ATN(X)	Arctangent of X is returned as an angle in radians in range +1.570796 to−1.570796
EXP(X)	Raises e to the power of X
LOG(X)	Returns the natural logarithm of X, $\ln(X)$. Logs to any other base must be calculated using the identity: $\log_A(X)$ = LOG(X) / LOG(A) where $\log_A(X)$ is the desired log to base A
SQR(X)	Returns the square root of X
ABS(X)	Returns the absolute value of X
SGN(X)	Returns 1, 0 or -1 to indicate the sign of X

Function calls can be used as expressions or elements of expressions wherever expressions are legal. The argument X of the function can be either a constant, a variable, an expression or another function.

Some useful mathematical functions which can be derived from standard BASIC functions are listed in Table A.2.

TABLE A.2 Derived Mathematical Functions

Function	Formula
TRIGONOMETRIC	
Cosecant	CSC(X)=1/SIN(X)
Cotangent	COT(X)=1/TAN(X)
Secant	SEC(X)=1/COS(X)
INVERSE TRIGONOMETRIC	
Arc Cosine	ACOS(X)=-ATN(X/SQR(-X*X+1))+PI/2
Arc Sine	ASIN(X)=ATN(X/SQR(-X*X+1))
Arc Cosecant	ACSC(X)=ATN(1/SQR(X*X-1))+(SGN(X)-1)*PI/2
Arc Cotangent	ACOT(X)=-ATN(X)+PI/2
Arc Secant	ASEC(X)=ATN(SQR(X*X-1))+(SGN(X)-1)*PI/2
HYPERBOLIC	
Hyp Cosine	COSH(X)=(EXP(X)+EXP(-X))/2
Hyp Sine	SINH(X)=(EXP(X)-EXP(-X))/2
Hyp Tangent	TANH(X)=-EXP(-X)/(EXP(X)+EXP(-X))*2+1
Hyp Cosecant	CSCH(X)=2/(EXP(X)-EXP(-X))
Hyp Cotangent	COTH(X)=EXP(-X)/(EXP(X)-EXP(-X))*2+1
Hyp Secant	SECH(X)=2/(EXP(X)+EXP(-X))
INVERSE HYPERBOLIC	
Arc Cosh	ACOSH(X)=LOG(X+SQR(X*X-1))
Arc Sinh	ASINH(X)=LOG(X+SQR(X*X+1))
Arc Tanh	ATANH(X)=LOG((1+X)/(1-X))/2
Arc Cosech	ACSCH(X)=LOG((SGN(X)*SQR(X*X+1)+1)/X)
Arc Cotanh	ACOTH(X)=LOG((X+1)/(X-1))/2
Arc Sech	ASECH(X)=LOG((SQR(-X*X+1)+1)/X)

Note: PI = 3.141592654 in above formulae.

Sketches of the hyperbolic functions and their definitions are shown in Fig.A.1.

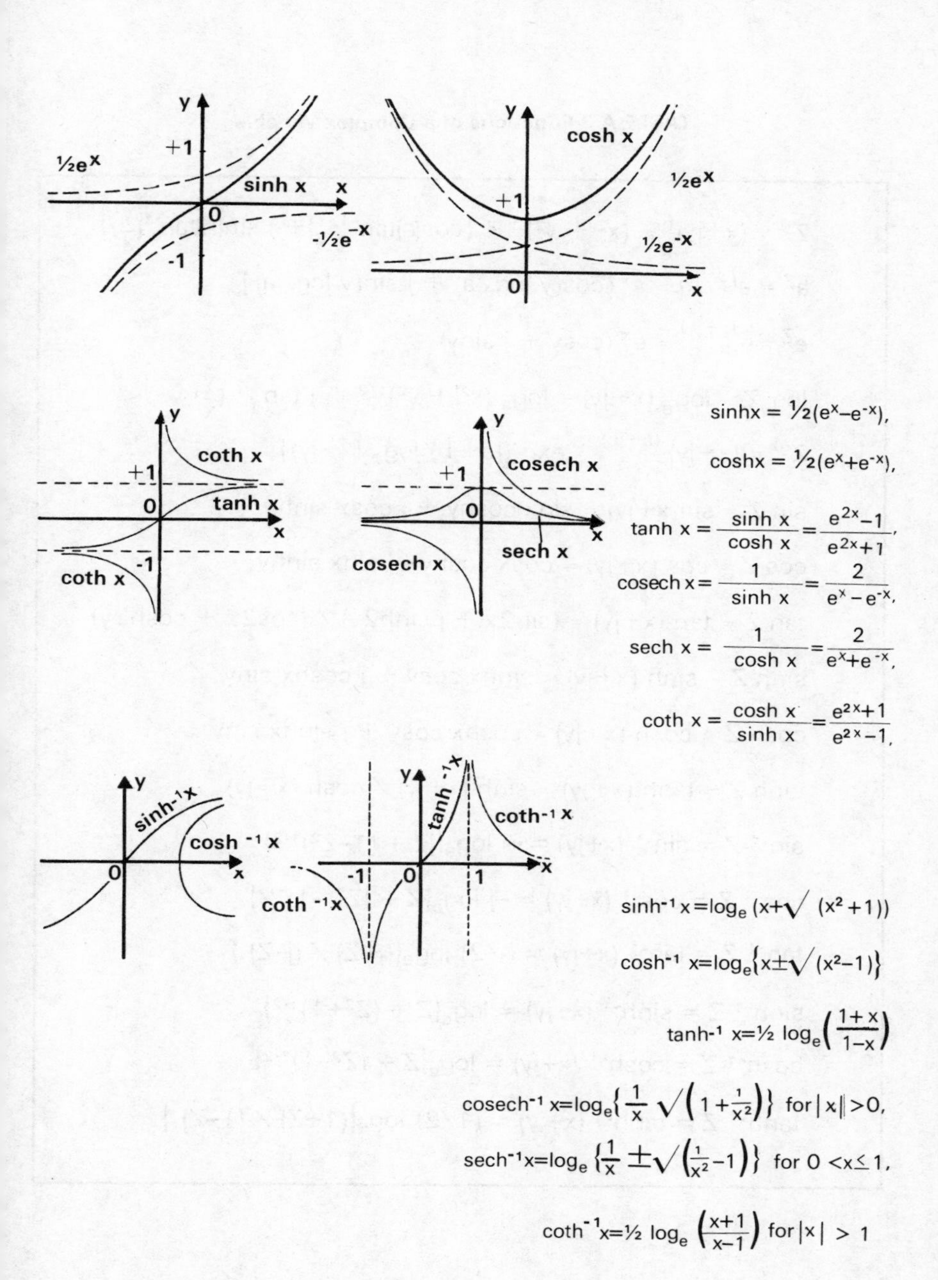

Figure A.1 Sketches of Hyperbolic functions and their definitions.

TABLE A.3 Functions of a Complex Variable

$$Z^n = (x+jy)^n = (x^2 + y^2)^{n/2} \{\cos(n)\tan^{-1}(\tfrac{y}{x}) + j \sin(n)\tan^{-1}(\tfrac{y}{x})\}$$

$$a^Z = a^{(x+jy)} = a^x\{\cos(y \log_e a) + j \sin(y \log_e a)\}$$

$$e^Z = e^{(x+jy)} = e^x (\cos y + j \sin y)$$

$$\log_e Z = \log_e (x+jy) = \log_e (x^2 + y^2)^{1/2} + j \tan^{-1} (\tfrac{y}{x})$$

$$Z^\omega = (x+jy)^{(a+jb)} = \exp \{(a+jb) \log_e (x+jy)\}$$

$$\sin Z = \sin(x+jy) = \sin x \cosh y + j \cos x \sinh y$$

$$\cos Z = \cos (x+jy) = \cos x \cosh y - j \sin x \sinh y$$

$$\tan Z = \tan (x+jy) = (\sin 2x + j \sinh 2y) / (\cos 2x + \cosh 2y)$$

$$\sinh Z = \sinh (x+jy) = \sinh x \cos y + j \cosh x \sin y$$

$$\cosh Z = \cosh (x+jy) = \cosh x \cos y + j \sinh x \sin y$$

$$\tanh Z = \tanh (x+jy) = \sinh (x+jy) / \cosh (x+jy)$$

$$\sin^{-1} Z = \sin^{-1} (x+jy) = -j \log_e\{jZ + (1-Z^2)^{1/2}\}$$

$$\cos^{-1} Z = \cos^{-1} (x+jy) = -j \log_e\{Z + (Z^2 - 1)^{1/2}\}$$

$$\tan^{-1} Z = \tan^{-1} (x+jy) = (j/2) \log_e\{(j+Z) / (j-Z)\}$$

$$\sinh^{-1} Z = \sinh^{-1} (x+jy) = \log_e\{Z + (Z^2+1)^{1/2}\}$$

$$\cosh^{-1} Z = \cosh^{-1} (x+jy) = \log_e\{Z + (Z^2-1)^{1/2}\}$$

$$\tanh^{-1} Z = \tanh^{-1} (x+jy) = (1/2) \log_e\{(1+Z) / (1-Z)\}$$

APPENDIX B

MICRO PROGRAMS

B.1 MICRO.ROOTS PROGRAM

The short BASIC program given in Listing B.1 can be used to find the roots of a single equation which must be typed into the computer as a BASIC expression in line 3. The first statement of line 7 incorporates the $\sinh^{-1}$ part of the new algorithm, while the rest of the algorithm is to be found in the second statement of line 8.

```
1 PRINT "ENTER EQUATION ON LINE 3 AS:-":
  PRINT "3 F=2*X-5":PRINT
2 INPUT "ACC=";D:INPUT "X=";X:INPUT "R=";R:INPUT "S=";S:
  P=0:S1=1:PRINT:PRINT "X"
7 H=S*LOG(ABS(F)+SQR(1+F*F))*SGN(F):S2=SGN(H):
  IF S1*S2>0 THEN P=P+1:R=R-1
8 R=R+1:X1=X+H*2↑(P/3-R-1/3):PRINT X:
  IF ABS(X-X1)<D THEN PRINT:PRINT "F=";F:END
9 X=X1:S1=S2:GOTO 3
```

Listing B.1. MICRO.ROOTS program for solving single equations.

The program shown above is written without the use of subroutines, in order to make it especially suitable for use with programmable calculators or hand-held computers which might not have this facility. In order to conserve memory space, the format of the program was changed slightly, though its accuracy remains unaltered. Line 1 is for information only and can be deleted.

B.2 MICRO.SIMROOTS PROGRAM

The very short program given below can solve single or up to 6 simultaneous equations, although this could be extended. The program requires minimum memory to operate and is suitable for a one line display pocket computers such as the SHARP 1500, with less than 1 kBytes of RAM, including the equations. In order to avoid the need for extra RAM units, the program has been severely trimmed and as a result the 'Fine Search' and 'Sign Search' routines are not included.

Equations are inserted into the program starting on line 10 in exactly the same format as that of the SIM.ROOTS program discussed in Section 3.2.

```
 1 PRINT "ENTER EQUATIONS ON LINES 10-15 AS:-":PRINT
   "10 F(0)=2*X(0)+X(1)-5:RETURN":PRINT
   "11 F(1)=X(0)+3*X(1)-2:RETURN"
 5 INPUT "NR OF EQUNS=";N1:INPUT "ACC=";D:INPUT "MAX
   STEPS=";M:
   INPUT "COMMON R=";R:N=N1:N1=N1-1:GOTO 22
22 DIM S(N),S1(N),S2(N),X(N),X1(N),X2(N),R(N),R1(N),
   P(N),H(N),F(N)
24 FOR I=0 TO N1:PRINT "X";I;"=";:INPUT "";X(I):X2(I)=X(I):
   R(I)=R:R1(I)=R(I):S1(I)=1:P(I)=0:NEXT I
26 IF N1=0 THEN INPUT "S=";S(0):PRINT:GOSUB 32:END
28 FOR I=0 TO N1:PRINT "S";I;"=";:INPUT "";S(I):NEXT I:PRINT
30 FOR I=0 TO N1:X(I)=X2(I):R(I)=R1(I):S1(I)=1:
   P(I)=0:NEXT I:GOSUB 32:GOTO 28
32 FOR K=1 TO M:L=-1:FOR I=0 TO N1:J=I+1:
   ON J GOSUB 10,11,12,13,14,15
33 IF ABS(X(I))>1E6 OR ABS(F(I))>1E18 THEN PRINT
   "X OR F ABOVE LIMIT SET IN LINE 33":K=M:NEXT K:RETURN
34 H(I)=S(I) * SGN(F(I)) * LOG(ABS(F(I))+SQR(F(I)*F(I)+1)):
   S2(I)=SGN(H(I))
36 IF S2(I)*S1(I) > 0 THEN P(I)=P(I)+1:R(I)=R(I)-1
42 R(I)=R(I)+1:X1(I)=X(I) + H(I) * 2↑(P(I)/3-R(I)-1/3):
```

```
   PRINT "X";I;"=";X1(I)
44 IF ABS(X1(I)-X(I)) < D THEN L=L+1
46 X(I)=X1(I):S1(I)=S2(I):NEXT I: IF L <> N1 GOTO 54
48 PRINT:PRINT K;" STEPS":FOR I=0 TO N1:PRINT "X";I;"=";X1(I);
50 IF L=N1 THEN J=I+1:ON J GOSUB 10,11,12,13,14,15
52 PRINT TAB(20);"F";I;"=";F(I):NEXT I:PRINT:RETURN
54 IF N1=0 THEN NEXT K:GOTO 58
56 PRINT:NEXT K
58 PRINT M;" STEPS, NO SOLN":RETURN
```

Listing B.2 MICRO.SIMS program for solving simultaneous equations.

Further memory could be saved by presetting the values for accuracy (D), maximum number of steps (M) and common R (R) in line 5.

APPENDIX C

SIM.ROOTS FOR THE BBC COMPUTER

The listing of the implemented SIM.ROOTS program for the BBC (model A or B) microcomputer is shown below. The program differs in very few, machine dependent, statements from the program discussed in the text. We show here the full listing as it can help in tracing the program.

```
  0 CLS:PRINT:PRINT "FOR ROOTS OF SINGLE EQU. USE
    LINE 10":PRINT
  1 PRINT "FOR SIMULTANEOUS EQUATIONS":
    PRINT "ENTER YOUR EQUATIONS AS FOLLOWS:":PRINT
  2 PRINT "10 F(0)=3 * X(0) -7 * X(1) + 5:RETURN"
  3 PRINT "11 F(1)=2 * X(0) + 3 * X(1) - 1:RETURN":PRINT
  7 PRINT TAB(5,23);"PRESS SPACE TO CONTINUE":PRINT TAB(3);
    OR <ESCAPE> TO ENTER EQUATIONS":T$=GET$:GOTO 100
100 DEF FNS(X) = LN(ABS(X)+SQR(X*X+1))
110 CLS
140 INPUT "NR OF EQUATIONS ",N:NN=N:N=N-1
145 INPUT "CONSTANTS (Y/N) ",T$:IF LEFT$(T$,1)="Y" THEN
    INPUT "HOW MANY ",G:DIM G(G):FOR I=0 TO G-1:
    PRINT "G(";I;")=";:INPUT G(I):NEXT I
150 DIM F(NN),X(NN),P(NN),S1(NN),H(NN),X1(NN),S(NN),
    S2(NN),X2(NN),R1(NN),A(NN),R(NN),A1(NN)
155 INPUT "DECIM ACCUR ",D:INPUT "NR OF ITERATIONS ",M
160 INPUT "OUTPUT RESULTS TO PRINTER (Y/N) ",TT$
170 FOR I=0 TO N:PRINT "X(";I;")=";:INPUT X(I):
    X2(I)=X(I):NEXT I
180 FOR I=0 TO N:PRINT "R(";I;")=";:INPUT R(I):
    R1(I)=R(I):NEXT I
190 FOR I=0 TO N:S1(I)=1:P(I)=0:A(I)=I:A1(I)=I:NEXT I
200 K1=0:K2=0:IF N=0 THEN INPUT "FINE SEARCH (Y/N) ",T$:
    IF LEFT$ (T$,1)="Y" THEN INPUT "SEARCH INCREMENT ",L:
    K1=1:K2=1
210 IF N=0 THEN INPUT "SIGN F(0) ",S(0):IF K1=1 THEN PRINT:
```

```
    PRINT "F(X)";TAB(25);"X":PRINT:GOSUB 325:GOTO 270
215 IF N=0 THEN GOSUB 320:GOTO 270
220 INPUT "RE ARRANGE EQUATIONS (Y/N) ",T$:
    IF LEFT$(T$,1)="N" GOTO 250
240 PRINT "ENTER EQUATION SEQUENCE":PRINT:
    PRINT "ORIGINAL F()", "NEW SEQUENCE":FOR I=0 TO N:
    PRINT I;TAB(23);: INPUT A(I):NEXT I
250 INPUT "SIGN SEARCH (Y/N) ",T$:IF LEFT$(T$,1)="Y"
    THEN GOSUB 600:GOTO 240
270 FOR I=0 TO N:PRINT "SIGN F(";I;")=";:INPUT S(I):
    NEXT I:FOR I2=0 TO N:I=A(I2):X(I)=X2(I):R(I)=R1(I2):
    S1(I)=1:P(I)=0:NEXT I2:GOSUB 300:IF N=0 GOTO 270
275 GOTO 220
300 PRINT:PRINT:PRINT "SIGN COMB.;
305 FOR I=0 TO N:PRINT TAB(3*I+12);S(I);:NEXT I:PRINT
310 PRINT:PRINT "EQU. SEQU. ";
315 FOR I=0 TO N:PRINT TAB(3*I+12);A(I);:NEXT I:PRINT
320 PRINT:PRINT "ITER";TAB(10);"P";TAB(17);"R";TAB(28);"ROOT":
    PRINT
325 FOR K=1 TO M:L1=-1:FOR I2=0 TO N:I=A(I2):I1=A1(I2)
330 J=I+1:ON J GOSUB 10,11,12,13,14,15,16,17,18,19,20,
    21,22,23,24,25,26,27,28,29,30,31,32,33,34,35,36,
    37,38,39,40,41,42,43,44,45
333 IF ABS(X(I1))>1E6 OR ABS(F(I))>1E18 THEN PRINT
    "X OR F ABOVE LIMIT SET IN LINE 333":K=M:NEXT K:RETURN
335 H(I)=S(I)*FNS(F(I))*SGN(F(I)):S2(I)=SGN(H(I))
340 IF K1=1 THEN GOSUB 505:IF K1=1 THEN GOSUB 10:GOTO 335
345 IF S2(I)*S1(I)>0 THEN P(I)=P(I)+1:R(I)=R(I)-1
400 R(I)=R(I)+1:X1(I1)=X(I1)+H(I)*2↑(P(I)/3-R(I)-1/3)
405 PRINT TAB(2);K;TAB(10);P(I);TAB(17);R(I);TAB(25);X1(I1)
410 IF ABS(X1(I1)-X(I1))<D THEN L1=L1+1
415 X(I1)=X1(I1):S1(I)=S2(I):NEXT I2:IF L1<>N GOTO 485
417 IF LEFT$(TT$,1)="Y" THEN PRINT CHR$(2)
420 PRINT:PRINT "SOLUTION";TAB(22);" RESIDUAL":PRINT
425 FOR I3=0 TO N:I=A1(I3):I1=A(I3)
430 PRINT "X";I;"=";X1(I);TAB(21);"F";I;"=";
435 IF L1=N THEN J=I+1:ON J GOSUB 10,11,12,13,14,15,16,
    17,18,19,20,21,22,23,24,25,26,27,28,29,30,
    31,32,33,34,35,36,37,38,39,40,41,42,43,44,45
440 PRINT TAB(25);F(I)
445 NEXT I3
450 PRINT:PRINT "SIGN COMB.";
455 FOR I=0 TO N:PRINT TAB(3*I+12);S(I);:NEXT I
460 PRINT:PRINT "EQU. SEQU.";
```

```
465 FOR I=0 TO N:PRINT TAB(3*I+12);A(I);:NEXT I:PRINT:PRINT
467 IF LEFT$(TT$,1)="Y" THEN PRINT CHR$(3)
470 PRINT "TO CONTINUE, PRESS SPACE BAR":
    PRINT "OR TO END, PRESS <ESCAPE>":T$=GET$:PRINT
480 PRINT:K=M:NEXT K:RETURN
485 SB$=INKEY$(0):IF SB$=" " THEN K=M:NEXT K:RETURN
490 IF N=0 THEN NEXT K:GOTO 500
495 PRINT:NEXT K
500 PRINT "NOT CONVERGING IN ";M;" ITERATIONS"
501 PRINT:PRINT "SIGN COMB.";
502 FOR I=0 TO N:PRINT TAB(3*I+12);S(I);:NEXT I
503 PRINT:PRINT "EQU. SEQU.";
504 FOR I=0 TO N:PRINT TAB(3*I+12);A(I);:NEXT I:PRINT:PRINT:
    PRINT "TO CONTINUE, TYPE ... CONT":END:RETURN
505 IF K2=1 THEN K2=0:S1(I)=S2(I):GOTO 515
510 PRINT F(I);TAB(25);X(I):IF S1(I)*S2(I)<=0 THEN K1=0:
    PRINT:PRINT "ITER";TAB(10);"P";TAB(17);"R";
    TAB(28);"ROOT":PRINT:RETURN
515 X(I)=X(I)+L:RETURN
600 T1=2↑NN-1:FOR T2=0 TO T1:T5=T2:FOR I2=0 TO N:
    I=A(I2):X(I)=X2(I):R(I)=R1(I2):S1(I)=1:P(I)=0:NEXT I2
610 FOR I=N TO 0 STEP-1:T3=T5/2:T4=INT(T3):
    IF T3-T4<0.001 THEN S(I)=(-1):GOTO 630
620 S(I)=1
630 T5=T4:NEXT I:GOSUB 300:NEXT T2
640 PRINT:PRINT "ALL SIGN COMBINATIONS EXHAUSTED":
    INPUT "RE-ARRANGE EQUATIONS (Y/N) ",T$:
    IF LEFT$(T$,1)="N" THEN PRINT:PRINT "END OF PROGRAM":END
650 RETURN
```

Listing C.1 Implementation of SIM.ROOTS on the BBC microcomputer.

APPENDIX D

LIST OF VARIABLES USED IN PROGRAMS

A list of variables or constants that should not be used in equations or in additional program lines in the main versions of the Roots program is given below. Instead you are advised to use other single-letter or single-letter followed by a single digit variables.

The line numbers in which program variables can be found are included in the lists below so that editing of these programs can be easily carried out by a user whose computer might not accept string variables and/or two letter variable names and who has only a screen editor.

ROOTS

D	110	250					
F	180	250	290				
H	180	210					
I	170	240	260				
K$	130						
K1	140	150	190	290			
K2	140	280					
L	140	300					
N	110	170					
P	120	200	210	240			
R	120	200	210	240	290		
S	100	150	180				
S1	150	200	260	280	290		
S2	180	200	260	280	290		
T$	7						
X	100	120	210	250	660	290	300
X1	210	240	250	260			

SIM.ROOTS

A 150 190 240 270 315 325 425 504 600
A1 150 190 325 425
D 155 410
D$ 417 467
F 150 333 335 440 510
G 145
H 150 335 400
I 145 170 180 190 240 270 305 315 325 330
  333 335 345 400 405 415 425 430 435 440
  502 504 505 510 515 600 610 620 630
I1 325 333 400 405 410 415 425
I2 270 325 415 600
I3 425 445
J 330 435
K 325 333 405 490 495
K1 200 210 340 510
K2 200 505
L 200 515
L1 325 410 415 435
M 155 325 333 500
N 140 170 180 190 200 210 215 240 270 305
  315 325 415 425 435 490 502 504 600 610
NN 140 150 600
P 150 190 270 345 400 405 600
R 150 180 270 345 400 405 510 600
R1 150 180 270 600
S 100 150 210 270 305 335 502 610 620
S1 150 190 270 345 415 505 510 600
S2 150 335 345 415 505 510
T$ 7 145 200 220 250 640
T1 600
T2 600 630
T3 610
T4 610 630
T5 600 610 630
T$ 160 417 467
X 100 150 170 270 333 400 410 415 510 515
  600
X1 150 400 405 410 415 430
X2 150 170 270 600

DIFF.ROOTS

```
 A      150  165  190  240  270  280  315  325  425  504
        600  860 1030 1300 1400 1510
 A1     150  165  190  325  425
 AA     165  705  770
 B       10   11   12   20   21   22   30   31   32   40
         41   42  110  725  745  900  940
 C       10   11   12   13   20   21   22   23   30   31
         32   33   40   41   42   43  705  720  810
 C7     110  285  840  850 1600
 C8     110  830  850 1600
 D      155  410
 D$     417  467  860 1700
 D0     110  130  705  710  720  730  760  770  800  810
        910 1020
 D1     110  120  130  160  230  250  260  300  310  400
        440  425  450  460  470  475
 F       10   11   12   13   20   21   22   23   30   31
         32   33   40   41   42   43  150  333  335  440
        510
 G      145
 H       10   11   12   13   20   21   22   23   30   31
         32   33   40   41   42   43  150  335  400  725
        800  820  830  850 1500 1600
 I      145  165  170  180  190  240  270  280  305  315
        325  330  333  335  345  400  405  415  425  430
        435  440  502  504  505  510  515  600  610  620
        630  705  720  725  730  755  760  770  810  860
        900  910  950 1200 1300 1310 1400 1410 1510
 I0     760  770 1200 1300 1400
 I1     325  333  400  405  410  415  425 1200
 I2     270  280  325  415  600
 I3     425  445
 II     710  720  725  730  735  740  760  765  770  775
        800  810  820  860  910  920  930 1300 1400 1510
 IJ     730  745  750  910  940  950
 J      330  435  735  740  750  760  765  770  775  800
        810  820  920  930  950 1010 1030 1100 1120 1140
       1160
 J0     425  440  760  770 1200 1210 1300 1310 1400 1410
 JJ     740  745  930  940
 K      325  333  405  480  485  490  495 1200
 K1     110  200  210  340  510
 K2     200  505
 KT     425  760 1010 1100 1120 1140 1160 1200 1300 1400
 L      200  515
 L1     325  410  415  435
 M      155  325  333  480  485  500
 N      110  140  165  170  180  190  200  210  215  240
        270  280  305  315  325  415  425  435  490  502
        504  600  610  700  705  760  770  860 1300 1400
       1510
```

```
ND     425  705  760 1010 1100 1120 1140 1160 1200 1300
      1400
NN     140  150  165  600  705
NQ    1020 1030
NX    1010 1030
NY    1010 1020
NZ    1020
P      150  190  270  280  345  400  405  600  860 1200
      1510
R      150  180  270  280  345  400  405  510  600  860
      1100 1140 1200 1510
R1     150  180  270  280  600  860 1100 1140 1510
S      100  150  165  210  270  305  335  502  610  620
      1100 1160 1300
S1     150  165  190  270  280  345  415  505  510  600
S2     150  335  345  415  505  510
SI     130  700  710  715  720  725
SS     110  120  130  700  705  710  730  760  770  800
       910
SS$    110  250  475
T      725  800  820  850 1640
T$       7  110  120  145  200  220  280  285  640  830
       840  850 1100 1110 1130 1150 1500 1710
T1     600
T2     600  630
T3     610
T4     610  630
T5     600  610  630
TT$    160  417  467  860 1700
UL     110  850 1600
X       10   11   12   13   20   21   22   23   30   31
        32   33   40   41   42   43  100  150  165  170
       270  280  333  400  410  415  510  515  600 1100
      1120
X1     150  400  405  410  415  430  810  860 1200 1510
X2     150  165  170  270  280  600  860 1100 1120 1510
```

APPENDIX E

UTILITY PROGRAMS

E.1 TURNING EQUATIONS INTO TEXT FILES

In order to avoid having to retype equations every time you want to solve them, you should turn them into text files which can be saved under appropriate names on the disk. Such a text file can then be EXECed into the ROOTS program at a later stage.

The following program will allow you to create text files of your equations on the APPLE II computer only. The BBC computer has a similar program incorporated in the disk operating system (see your operating manual for instructions on how to use the command *SPOOL).

```
0    INPUT"NAME TO SAVE UNDER?";X$:D$=CHR$(4):?D$;"OPEN";X$:
     ?D$;"DELETE";X$:?D$;"OPEN";X$:?D$;"WRITE";X$:POKE 33,30:
     LIST 1,63999:?D$;"CLOSE";X$:POKE 33,40:?"DONE":END

1    INPUT"NAME TO SAVE UNDER?";X$:D$=CHR$(4):?D$;"OPEN";X$:
     ?D$;"DELETE";X$:?D$;"OPEN";X$:?D$;"WRITE";X$:POKE 33,30:
     LIST 2,63999:?D$;"CLOSE";X$:POKE 33,40:?"DONE":END
```

Note that the only difference between the above two program-lines is in the statement LIST. On RUNning the program, you will be asked:

NAME TO SAVE UNDER?

Type CAPTURE which will cause the second line of the program (line 1) to be turned into a text file named CAPTURE.

To turn equations into text files, DELete all unwanted program lines from memory, insert the disk containing the program CAPTURE and type:

EXEC CAPTURE

and when the 'in use' light of the disk drive goes off, type RUN.
On supplying a file name in response to the computer request, your equation(s) will be saved on disk in text-file form.

To incorporate such equation(s) into any version of the Roots program, LOAD the Roots program, insert the disk containing the equation(s) in text-file form and type EXEC 'filename', where 'filename' is the name of the equation(s) in text-file form. The equations will be merged with the Roots program in the correct position within it.

WARNING: In the case of the DIFF.ROOTS program, only the problem and its derived equations should be captured in this way. It is very important that you DELete the Taylor's series equations from the program in memory prior to RUNning the CAPTURE program. If you don't, the Taylor's equations will be also partly captured ('partly', because of their extreme length). EXECking such a text file into the DIFF.ROOTS program will overwrite the Taylor's equations which will cause program failure.

E.2 FUNCTION PLOTTER

The following program allows you to plot a single variable function, either on the screen or on the printer. The function must be typed in line 10, as in the example shown below:

```
 0 HOME:PRINT TAB(10);"*** FUNCTION PLOTTER ***":PRINT
 2 INPUT "OUTPUT TO PRINTER? (Y/N) ";P$:IF LEFT$(P$,1)="Y"
   THEN PRINT CHR$(4);"PR#1":REM CONNECT TO PRINTER
 4 INPUT "PLOT WIDTH? ";W:PRINT "X-AXIS POSITION? (0 - ";
   W-3;")";:INPUT " ";P
 6 INPUT "F-MULTIPLIER? ";S:INPUT "F-SCALE SHIFT? ";A
 8 INPUT "STARTING VALUE OF X? ";X:INPUT "INCREMENT VALUE
   OF X? ";I:PRINT:PRINT "!++++! = ";5/S;" UNITS":PRINT:
   PRINT SPC(INT(W-15));"X=";X:GOSUB 50:N=0:GOTO 40
10 F=15*SIN(X*3.14159265/180)
15 F=F+A:IF N=20 THEN N=0:GOSUB 50:GOTO 40
20 IF N=10 THEN GOSUB 45:GOTO 40
25 Z=INT(F*S+.5):Y=Z+P-A:IF A<Z THEN GOSUB 55:GOTO 40
30 IF A>Z THEN GOSUB 65:GOTO 40
35 PRINT SPC(Y);"*"
40 X=X+I:N=N+1:GOTO 10
45 FOR J=1 TO (INT(W/5)-3):PRINT "!++++";:NEXT J:
   PRINT "X=";X:RETURN
50 FOR J=1 TO (INT(W/5)-1):PRINT "!++++";:NEXT J:
   PRINT "!+++":RETURN
```

```
55 IF Z>(W-2-P+A) THEN PRINT SPC(P);"!";SPC(W-3-P);">":RETURN
60 PRINT SPC(P);"!";SPC(Y-P-1);"*":RETURN
65 IF Z<(-P+A) THEN PRINT "<";:IF P>0 THEN PRINT
   SPC(P-1);"!":RETURN
70 IF Z<(-P+A) THEN PRINT:RETURN
75 PRINT SPC(Y);"*";SPC(P-Y-1);"!":RETURN
```

The above program can be easily converted to use the PRINT TAB in place of the PRINT SPC function (in lines 35, 55, 60, 65 & 75) by observing the following:

(a) If your computer supports TAB(0) (e.g. the BBC) then replace all first occurrences of SPC by TAB and all second occurrences of SPC by TAB with an argument equal to the sum of the arguments of the first and second SPC functions in each effected line.

(b) If your computer does not support TAB(0) (e.g. the Apple II), then proceed as in (a) above, but also replace all occurrences of P and Y by (P+1) and (Y+1).

You can calibrate the system by trying a function such as F=10 which should give a vertical line of asterisks.

The following example which uses the sine function listed in line 10, is used to illustrate the program. If you modify the program, check that you get the same results with the same input parameters.

On the Apple II, plotting can be stopped by pressing the RESET key or temporarily halted by pressing CTRL'S' (restarting execution by another CTRL'S'). On the BBC, plotting can be stopped by pressing the ESCAPE key or halted by the CTRL'SHIFT' key.

```
OUTPUT TO PRINTER? (Y/N) Y
PLOT WIDTH? 65
X-AXIS POSITION? (0-62) 55
F-MULTIPLIER? 2
F-SCALE SHIFT?-30
STARTING VALUE OF X? 0
INCREMENT VALUE OF X? 10
```

```
!++++! = 2.5 UNITS
                                                 X=0
!++++!++++!++++!++++!++++!++++!++++!++++!++++!++++!++++!++++!+++
                              *                        !
                                   *                   !
                                        *              !
                                            *          !
                                                *      !
                                                   *   !
                                                     * !
                                                       *
                                                       *
!++++!++++!++++!++++!++++!++++!++++!++++!++++!++++X=100
                                                     * !
                                                   *   !
                                                *      !
                                            *          !
                                        *              !
                                   *                   !
                              *                        !
                         *                             !
                    *                                  !
!++++!++++!++++!++++!++++!++++!++++!++++!++++!++++!++++!++++!+++
          *                                            !
      *                                                !
   *                                                   !
<                                                      !
<                                                      !
<                                                      !
<                                                      !
<                                                      !
<                                                      !
 !++++!++++!++++!++++!++++!++++!++++!++++!++++!++++X=300
   *                                                   !
       *                                               !
           *                                           !
                *                                      !
                     *                                 !
                          *                            !
```

Note that by allowing the function to be plotted in a chart form (X in the vertical direction), we can examine the function behaviour from any starting value of X. Roots coincide with the vertical axis (!) only if the F-scale Shift is equal to 0. The F-scale Shift allows plotting to start in any position with respect to the X-axis position. Joint manipulation of the F-Scale Multiplier, F-scale Shift and X-axis position, allows you to blow up part of a function and yet retain it within your viewing window.

By using extra lines, you can operate on a function to

(a) compress it, e.g.

```
11 F = ASINH(F) or
11 F = ATAN(F) or
11 F = LOG(F)
```

(for ASINH() see Table A.2),

(b) expand it, e.g.

11 F = F^3.

With this program, you can now plot outputs from any of the ROOTS programs, especially those of differential equations.

E.3 SIMPSON'S RULE

The following program will integrate an explicit function, which must be typed in line 10.

RUN the program and provide the initial parameters requested which are the 'lower' and 'upper' limits of integration and the decimal accuracy to within which you require the answer to be evaluated.

```
 5 HOME:PRINT "INTEGRATOR – ENTER FUNCTION AS:–":PRINT:
   PRINT "10 F = 0.5*X + 3.34:RETURN":PRINT:GOTO 20
10 F = 3.34779568+2.38848736*X–2.18547104*X↑2
   +.074258824*X↑4:RETURN
20 INPUT "LOWER LIMIT ";A:INPUT "UPPER LIMIT ";B:
   INPUT "DECIM ACCUR ";D
30 H=(B–A)/2:X=A:GOSUB 10:J=F:X=B:GOSUB 10:J=H*(J+F):
   I1=3*J:N=1:PRINT:PRINT "DIVISOR","INTEGRAL":PRINT
40 S=0:FOR K=1 TO N:X=A+(2*K–1)*H:GOSUB 10:S=S+F:NEXT K
```

```
50 I0=J+4*H*S:IF ABS(I0-I1)<D THEN PRINT: GOTO 70
60 PRINT N,I0/3:J=(J+I0)/4:N=2*N:H=H/2:I1=I0:GOTO 40
70 PRINT "NR OF INCREMENTS=";2*N:PRINT "INCREMENT H=";H:
   I=I0/3:PRINT "INTEGRAL=";I:END
```

The function appearing to the right of the equals sign of line 10 is the fitted polynomial of Section 6.1. On RUNning the program and providing the initial parameters, the correct answer, within the specified accuracy, is reached after the calculation of 128 equally-spaced ordinates in the x-axis, as shown below:

```
INTEGRATOR – ENTER FUNCTION AS:–

10 F = 0.5*X + 3.34:RETURN

LOWER LIMIT .1
UPPER LIMIT 2.2
DECIM ACCUR .000001

DIVISOR          INTEGRAL

  1            5.83300971
  2            5.80931592
  4            5.80783505
  8            5.8077425
 16            5.80773672
 32            5.80773636

NR OF INCREMENTS=128
INCREMENT H=.01640625
INTEGRAL=5.80773633
```

INCREMENT H is automatically derived by the computer program in order to attain the specified accuracy.

E.4 EVALUATING DETERMINANTS

Amongst the easiest methods of solving a small number of linear simultaneous equations is by the determinant technique. The method is best explained with an example.

To solve the following three simultaneous equations:

$$2x_0 + x_1 - 2x_2 = -6$$
$$x_0 + x_1 + x_2 = 2$$
$$-x_0 - 2x_1 + 3x_2 = 12$$

for x_0, x_1, and x_2, evaluate the determinant D, formed by the coefficients a_{ij} of the three equations, i.e.,

$$D = \begin{vmatrix} 2 & 1 & -2 \\ 1 & 1 & 1 \\ -1 & -2 & 3 \end{vmatrix} = 8$$

To find the value of x_0, replace the coefficients a_{i0} of x_{i0} by the right-hand sides b_i, of the equations and evaluate the resultant determinant

$$P = \begin{vmatrix} -6 & 1 & -2 \\ 2 & 1 & 1 \\ 12 & -2 & 3 \end{vmatrix} = 8$$

The value of x_0 is then obtained by the ratio P/D.

Similarly, evaluate the replacement determinants

$$Q = \begin{vmatrix} 2 & -6 & -2 \\ 1 & 2 & 1 \\ -1 & 12 & 3 \end{vmatrix} = -16$$

and

$$R = \begin{vmatrix} 2 & 1 & -6 \\ 1 & 1 & 2 \\ -1 & -2 & 12 \end{vmatrix} = 24$$

to obtain x_1 =Q/D and x_2 =R/D.

The following program will evaluate the determinant of N number of simultaneous equations.

```
 0 HOME:PRINT "DETERMINANT EVALUATOR":PRINT
 2 INPUT "N=";N=N-1:DIM A(N,N)
 4 FOR I=0 TO N:FOR J=0 TO N:PRINT "A(";I;",";J;")=";:
   INPUT "";A(I,J):NEXT J:NEXT I
 6 FOR M=N TO 1 STEP-1:P=A(M,M):IF P=0 THEN PRINT "ERROR":END
 8 FOR I=0 TO M-1:Q=A(I,M)/P:FOR J=0 TO M:
   A(I,J)=A(I,J)-Q*A(M,J):NEXT J:NEXT I:NEXT M
10 D=A(0,0):FOR I=1 TO N:D=D*A(I,I):NEXT I:PRINT: PRINT "DET
   D=";D:END
```

As an example RUN the program and supply the coefficients a_{ij} of the above equations, where i refers to the row number and j to the column number of the coefficients.

APPENDIX F

ADDITIONS TO SIM.ROOTS TO IMPLEMENT DIFF.ROOTS

Adding the following statements to SIM.ROOTS will result in the final version of the program which we shall call DIFF.ROOTS.
Lines 4-43 and 700-1720 are new additions, while lines 110-475 could exist in the SIM.ROOTS version and only require some minor additions.

For the BBC computer, replace the semicolons following the delimited part of the INPUT statements by a comma.

```
 4 PRINT "FOR DIFFERENTIAL EQUATIONS":PRINT "ENTER
   EQUATIONS AS FOLLOWS:":PRINT:PRINT "FOR 0TH ORDER:-
   START ON LINE 10":
   PRINT "FOR 1ST ORDER:- START ON LINE 11"
 5 PRINT "FOR 2ND ORDER:-START ON LINE 12":PRINT "FOR
   3RD ORDER:- START ON LINE 13":PRINT "FOR 4TH ORDER:-
   START ON LINE 14": PRINT
 6 PRINT "FOR SIM. DIFF. EQU. USE LINES STARTING":
   PRINT "AT 10, 20, 30 & 40 FOR 1, 2, 3 & 4 SETS":PRINT
10 F(0)=X(0) - (C(0) + C(1)*H*(1-B(3)) + C(2)*H↑2/2*(1-B(2))
   + C(3)*H↑3/6*(1-B(1)) + H*X(1)*B(3) + H↑2/2*X(2)*B(2)
   + H↑3/6*X(3)*B(1) + H↑4/24*X(4) + H↑5/120*X(5)):RETURN
11 F(1)=X(1) - (C(1) + C(2)*H*(1-B(2)) + C(3)*H↑2/2*(1-B(1))
   + H*X(2)*B(2) + H↑2/2*X(3)*B(1) + H↑3/6*X(4)
   + H↑4/24*X(5)):RETURN
12 F(2)=X(2)- (C(2) + C(3)*H*(1-B(1)) + H*X(3)*B(1)
   + H↑2/2*X(4) + H↑3/6*X(5)):RETURN
13 F(3)=X(3)- (C(3) + H*X(4) + H↑2/2*X(5)):RETURN
20 F(10)=X(10)- (C(10) + C(11)*H*(1-B(6)) + C(12)*H↑2/2*(1-
   B(5))
```

```
    + C(13)*H↑3/6*(1-B(4)) + H*X(11)*B(6) + H↑2/2*X(12)*B(5)
    + H↑3/6*X(13)*B(4) + H↑4/24*X(14) + H↑5/120*X(15))
    :RETURN
 21 F(11)=X(11) - (C(11) + C(12)*H*(1-B(5)) + C(13)*H↑2/2*(1-
    B(4))
    + H*X(12)*B(5) + H↑2/2*X(13)*B(4) + H↑3/6*X(14)
    + H↑4/24*X(15)):RETURN
 22 F(12)=X(12) - (C(12) + C(13)*H*(1-B(4)) + H*X(13)*B(4)
    + H↑2/2*X(14) + H↑3/6*X(15)):RETURN
 23 F(13)=X(13) - (C(13) + H*X(14) + H↑2/2*X(15)):RETURN
 30 F(20)=X(20) - (C(20) + C(21)*H*(1-B(9)) + C(22)*H↑2/2*(1-
    B(8))
    + C(23)*H↑3/6*(1-B(7)) + H*X(21)*B(9) + H↑2/2*X(22)*B(8)
    + H↑3/6*X(23)*B(7) + H↑4/24*X(24) + H↑5/120*X(25)):
    RETURN
 31 F(21)=X(21) - (C(21) + C(22)*H*(1-B(8)) + C(23)*H↑2/2*(1-
    B(7))
    + H*X(22)*B(8) + H↑2/2*X(23)*B(7) + H↑3/6*X(24)
    + H↑4/24*X(25)):RETURN
 32 F(22)=X(22)- (C(22) + C(23)*H*(1-B(7)) + H*X(23)*B(7)
    + H↑2/2*X(24) + H↑3/6*X(25)):RETURN
 33 F(23)=X(23) - (C(23) + H*X(24) + H↑2/2*X(25)):RETURN
 40 F(30)=X(30) - (C(30) + C(31)*H*(1-B(12)) + C(32)*H↑2/2*
    (1-B(11)) + C(33)*H↑3/6*(1-B(10)) + H*X(31)*B(12) + H↑2/2*
    X(32)*B(11) + H↑3/6*X(33)*B(10) + H↑4/24*X(34) + H↑5/120*
    X(35)):RETURN
 41 F(31)=X(31)- (C(31) + C(32)*H*(1-B(11)) + C(33)*H↑2/2*
    (1-B(10)) + H*X(32)*B(11) + H↑2/2*X(33)*B(10) + H↑3/6*X(34)
    + H↑4/24*X(35)):RETURN
 42 F(32)=X(32)- (C(32) + C(33)*H*(1-B(10)) + H*X(33)*B(10)
    + H↑2/2*X(34) + H↑3/6*X(35)):RETURN
 43 F(33)=X(33) - (C(33) + H*X(34) + H↑2/2*X(35)):RETURN
110 HOME:D1=0:INPUT "DIFFERENTIAL EQUATIONS? (Y/N) ";T$:
    IF LEFT$(T$,1)="Y" THEN D1=1:SS=0:K1=0:SS$=" ":C7=0:C8=0:
    UL=0:DIM B(12),D0(9),N(9)
120 IF D1=1 THEN INPUT "SIMULTANEOUS? (Y/N) ";T$:IF LEFT$(T$,1)=
    "Y" THEN INPUT "HOW MANY? (2/3/4) ";SS:IF SS<2 OR SS>4
    GOTO 120
130 IF D1=1 THEN PRINT:FOR SI=1 TO SS:PRINT "DIFF ORDER IN
    SUB-SET ";SI;:INPUT "? (0/1/2/3/4) ";D0(SI):NEXT SI:
    GOSUB 700:GOTO 145
160 INPUT "OUTPUT RESULTS TO PRINTER? (Y/N) "; TT$:
    IF D1=0 GOTO 170
165 NN=NN-5:FOR I=0 TO NN:S1(I)=1:S(I)=-1:X(I)=0:X2(I)=X(I):
    NEXT I:FOR I=0 TO N:A1(I)=AA(I):A(I)=AA(I):NEXT I:GOTO 220
```

```
230 IF D1=1 THEN GOSUB 1000:GOTO 260
250 IF D1=0 THEN INPUT "SIGN SEARCH? (Y/N) ";SS$:IF LEFT$
    (SS$,1)="Y" THEN GOSUB 600:GOTO 240
260 IF D1=1 THEN GOSUB 1100:GOSUB 300:GOTO 280
280 FOR I2=0 TO N:I=A(I2):X(I)=X2(I):R(I)=R1(I2):S1(I)=1:
    P(I)=0:NEXT I2:GOSUB 800:IF LEFT$(T$,1)="S" THEN
    GOSUB 1500:GOTO 220
285 IF C7=0 THEN PRINT:IF LEFT$(T$,1)="C" THEN GOSUB 1600
290 GOSUB 300:GOTO 280
300 PRINT:PRINT:PRINT "SIGN COMB.;:IF D1=1 THEN GOSUB
    1300: GOTO 310
310 PRINT:PRINT "EQU. SEQU.;:IF D1=1 THEN GOSUB 1400:GOTO
    320
400 R(I)=R(I)+1:X1(I1)=X(I1) + H(I)*2 ↑ (P(I)/3 - R(I) - 1/3):
    IF D1=1 THEN GOSUB 1200:GOTO 410
425 FOR I3=0 TO N:I=A1(I3):I1=A(I3):IF D1=1 THEN FOR J0=1 TO
    KT: IF I<>ND(J0) THEN NEXT J0: GOTO 445
440 PRINT TAB(25);F(I):IF D1=1 THEN NEXT J0
450 PRINT:PRINT "SIGN COMB.;:IF D1=1 THEN GOSUB 1300:GOTO
    460
460 PRINT:PRINT "EQU. SEQU.;:IF D1=1 THEN GOSUB 1400:GOTO
    470
470 IF D1=0 THEN PRINT "TO CONTINUE, TYPE ... CONT":
    PRINT:PRINT:END
475 IF D1=1 AND LEFT$(SS$,1)="Y" THEN GOSUB 800
501 PRINT:PRINT "SIGN COMB.;:IF D1=1 THEN GOSUB 1300:GOTO
    503
503 PRINT:PRINT "EQU. SEQU.;:IF D1=1 THEN GOSUB 1400:RETURN
700 FOR SI=1 TO SS:PRINT "NR OF EQU IN SUB-SET ";SI;:
    INPUT "? ";N(SI):NEXT SI
705 FOR I=1 TO SS:NN=4 + N(I) + D0(I) + 10*(I-1):NEXT I:
    DIM C(NN),ND(NN),AA(NN)
710 PRINT "ENTER INITIAL VALUES":II=0:FOR SI=1 TO SS:
    IF D0(SI)=0 GOTO 725
715 PRINT "FOR SUB-SET ";SI
720 FOR I=0 TO D0(SI)-1:PRINT "C(";I+II;")";:
    INPUT "=";C(I+II):NEXT I
725 II=II+10:NEXT SI:INPUT "H=";H:INPUT "T=";T:FOR I=1 TO 12:
    B(I)=0:NEXT I
730 IJ=0:FOR I=1 TO SS:II=D0(I):IF II=0 GOTO 755
735 FOR J=4 TO II STEP-1
740 JJ=J-II:IF JJ=0 GOTO 750
745 B(JJ+IJ)=1
750 NEXT J:IJ=IJ+3
755 NEXT I
```

```
760 II=0:KT=0:FOR J=1 TO SS:J0=N(J):I0=D0(J):FOR I=I0 TO J0+I0-
    1: KT=KT+1:ND(KT)=I+II:NEXT I
765 II=II+10:NEXT J
770 II=0:N=-1:FOR J=1 TO SS:J0=N(J):I0=D0(J):FOR I=0 TO J0+I0-
    1: N=N+1:AA(N)=I+II:NEXT I
775 II=II+10:NEXT J:RETURN
800 T=T+H:PRINT:II=0:FOR J=1 TO SS:IF D0(J)=0 GOTO 820
810 FOR I=0 TO D0(J)-1:C(I+II)=X1(I+II):PRINT "C(";I+II;")=";
    C(I+II):NEXT I
820 II=II+10:NEXT J:PRINT "H=";H:PRINT "T=";T:PRINT
830 IF C8=0 THEN GOSUB 1700:IF LEFT$(T$,1)="Y" THEN
    INPUT "H = ";H:GOTO 860
840 IF LEFT$(T$,1)<>"C" THEN C7=0:GOTO 860
850 IF ABS(T-UL)<0.000001 THEN C7=0:C8=0:GOSUB 1700:
    IF LEFT$(T$,1)="Y" THEN INPUT "H = ";H:RETURN
860 FOR II=0 TO N:I=A(II):P(I)=0:R(I)=R1(I):X2(I)=X1(I):
    NEXT II:IF LEFT$(TT$,1)="Y" THEN PRINT D$;"PR#0"
870 RETURN
900 FOR I=1 TO 12:B(I)=0:NEXT I
910 IJ=0:FOR I=1 TO SS:II=D0(I)
920 FOR J=4 TO II STEP-1
930 JJ=J-II:IF JJ=0 GOTO 950
940 B(JJ+IJ)=1
950 NEXT J:IJ=IJ+3:NEXT I:RETURN
1000 PRINT "ENTER EQUATION SEQUENCE":PRINT:PRINT "ORIGINAL
     F()", "NEW SEQUENCE"
1010 FOR J=1 TO KT:PRINT ND(J);TAB(23);:INPUT "";NX:
     NY=INT(NX/10)+1
1020 NQ=0:FOR NZ=1 TO NY:NQ=NQ+D0(NZ):NEXT NZ
1030 A(J+NQ-1)=NX:NEXT J:RETURN
1100 FOR J=1 TO KT:X(ND(J))=0:X2(ND(J))=0:R(ND(J))=0:
     R1(ND(J))=0:S(ND(J))=-1:NEXT J:PRINT:
     PRINT "DEFAULT STARTING PARAMETERS ARE:":
     PRINT "X()=0, R()=0, SIGN()=-1 ";:INPUT "(Y/N) ";T$:
     IF LEFT$(T$,1)<>"N" THEN RETURN
1110 INPUT "CHANGE X()? (Y/N) ";T$:IF LEFT$(T$,1)="N" GOTO
     1130
1120 FOR J=1 TO KT:PRINT "X(";ND(J);")";:INPUT " = ";X(ND(J)):
     X2(ND(J))=X(ND(J)):NEXT J
1130 INPUT "CHANGE R()? (Y/N) ";T$:IF LEFT$(T$,1)="N" GOTO
     1150
1140 FOR J=1 TO KT:PRINT "R(";ND(J);")";:INPUT " = ";R(ND(J)):
     R1(ND(J))=R(ND(J)):NEXT J
1150 INPUT "CHANGE SIGNS? (Y/N) ";T$:IF LEFT$(T$,1)="N"
     THEN RETURN
```

```
1160 FOR J=1 TO KT:PRINT "SIGN F(";ND(J);")";:
     INPUT " = ";S(ND(J)):NEXT J:RETURN
1200 FOR J0=1 TO KT:I0=ND(J0):IF I0=I THEN PRINT TAB(2);K;
     TAB(10);P(I);TAB(17);R(I);TAB(25);X1(I1)
1210 NEXT J0:RETURN
1300 J1=(-1):FOR I=0 TO N:II=A(I):FOR J0=1 TO KT:I1300 J1=(-
     1):FOR I=0 TO N:II=A(I):FOR J0=1 TO KT:I0=ND(J0):
     IF I0=II THEN J1=J1+1:PRINT TAB(3*J1+12);S(II);
1310 NEXT J0:NEXT I:PRINT:RETURN
1400 J1=(-1):FOR I=0 TO N:II=A(I):FOR J0=1 TO KT:I0=ND(J0):
     IF I0=II THEN J1=J1+1:PRINT TAB(3*J1+12);A(I);
1410 NEXT J0:NEXT I:PRINT:RETURN
1500 INPUT "RE-SET H? (Y/N) ";T$:IF LEFT$(T$,1)="Y"
     THEN INPUT "H = ";H
1510 FOR II=0 TO N:I=A(II):P(I)=0:R(I)=R1(I):X2(I)=X1(I):
     NEXT II:RETURN
1600 INPUT "NEXT LIMIT OF T? ";UL:INPUT "NEW VALUE OF H? ";H:
     C7=1:C8=1:RETURN
1700 IF LEFT$(TT$,1)="Y" THEN PRINT D$;"PR#0"
1710 INPUT "CHANGE H? (Y/N/S/C/Q) ";T$:IF LEFT$(T$,1)="Q"
     THEN END
1720 RETURN
```

Listing F.1 Requisite additions to SIM.ROOTS to form the DIFF.ROOTS program.

BIBLIOGRAPHY

The following texts, although not amongst the latest we have consulted in the field of numerical analysis, are, nevertheless, an excellent introduction to the subject.

Balfour, A. and McTernan, A., "Numerical Solution of Equations", Heinemann Educational Books Ltd., London, 1967.

Bartlett, B. R. and Fyfe, D. J., "Handbook of Mathematical Formulae for Engineers and Scientists", Denny Publications Ltd., London, 1974.

Cohen, A. M., "Numerical Analysis", McGraw-Hill, London, 1973.

Conte, S. D. and de Boor, C., "Elementary Numerical Analysis", 2nd Edition, McGraw-Hill, Tokyo, 1972.

Gerald, C. F., "Applied Numerical Analysis", Addison-Wesley, Philippines, 1970.

Noble, B., "Numerical Methods: 1 Iteration Programming and Algebraic Equations", Oliver & Boyd, Edinburgh, 1970.

Noble, B., "Numerical Methods: 2 Differences, Integration and Differential Equations", Oliver & Boyd, Edinburgh, 1972.

Lamb, H., "An Elementary Course of Infinitesimal Calculus", 3rd Edition, Cambridge University Press, 1956.

Shampine, L. F. and Allen, R. C., "Numerical Computing: An Introduction", W. B. Saunders Company, Philadelphia, 1973.

Stanton, R. G., "Numerical Methods for Science and Engineering", Prentice-Hall, New Jersey, 1961.

Watson, W. A., Philipson, T. and Oates, P. J., "Numerical Analysis - The Mathematics of Computing 1", Edward Arnold, London, 1972.

Watson, W. A., Philipson, T. and Oates, P. J., "Numerical Analysis - The Mathematics of Computing 2", Edward Arnold, London, 1974.